WORKSHEETS FOR CLASSROOM OR LAB PRACTICE

CARRIE GREEN

INTEGRATED ARITHMETIC AND BASIC ALGEBRA

FIFTH EDITION

Bill E. Jordan
Seminole State College

William P. Palow
Miami-Dade College

PEARSON

Boston   Columbus   Indianapolis   New York   San Francisco   Upper Saddle River
Amsterdam   Cape Town   Dubai   London   Madrid   Milan   Munich   Paris   Montreal   Toronto
Delhi   Mexico City   Sao Paulo   Sydney   Hong Kong   Seoul   Singapore   Taipei   Tokyo

The author and publisher of this book have used their best efforts in preparing this book. These efforts include the development, research, and testing of the theories and programs to determine their effectiveness. The author and publisher make no warranty of any kind, expressed or implied, with regard to these programs or the documentation contained in this book. The author and publisher shall not be liable in any event for incidental or consequential damages in connection with, or arising out of, the furnishing, performance, or use of these programs.

Reproduced by Pearson from electronic files supplied by the author.

Copyright © 2013, 2009, 2005 Pearson Education, Inc.
Publishing as Pearson, 75 Arlington Street, Boston, MA 02116.

All rights reserved. No part of this publication may be reproduced, stored in a retrieval system, or transmitted, in any form or by any means, electronic, mechanical, photocopying, recording, or otherwise, without the prior written permission of the publisher. Printed in the United States of America.

ISBN-13: 978-0-321-75924-5
ISBN-10: 0-321-75924-9

8 16

www.pearsonhighered.com

PEARSON

Table of Contents

Chapter R..1

Chapter 1..35

Chapter 2..71

Chapter 3..113

Chapter 4..157

Chapter 5..221

Chapter 6..253

Chapter 7..281

Chapter 8..317

Chapter 9..341

Chapter 10..377

Chapter 11..415

Odd Answers...443

Copyright © 2013 Pearson Education, Inc.

Name: Date:
Instructor: Section:

Chapter R   BASIC IDEAS

R.1   Reading and Writing Numerals

Learning Objectives
A   Read and write numerals that represent whole numbers.
B   Write whole numbers in expanded notation.
C   Write the word name for a standard numeral.

Key Terms

Use the vocabulary terms listed below to complete each statement in exercises 1–5.

place value   word name   numeral   digits   expanded notation

1. 0,1,2,3,4,5,6,7,8,9 are symbols that are called _____.

2. A(n) _____ is a symbol that represents a number.

3. The specific location of a digit is called the _____.

4. $6000 + 200 + 80 + 4$ is the numeral 6284 written in _____.

5. Six thousand, two hundred eighty-four is the _____ for 6284.

Practice Problems

Objective A   Read and write numerals that represent whole numbers.

Give the place value of the underlined digit in each of the following.

1. 97,6<u>4</u>1   1. _____

2. <u>6</u>354   2. _____

3. <u>6</u>,214,539   3. _____

4. 8<u>2</u>1,067,935   4. _____

5. 6<u>2</u>   5. _____

Copyright © 2013 Pearson Education, Inc.

6. 1<u>4</u>1,036   6._____

What is the meaning of the underlined digit in each of the following numerals?

7. 17,<u>8</u>45   7._____

8. 43,6<u>9</u>1,076   8._____

9. <u>1</u>578   9._____

10. <u>2</u>35,691   10._____

Objective B Write whole numbers in expanded notation.

Write each of the following in expanded form.

11. 263   11._____

12. 468,932   12._____

13. 1,471,265   13._____

14. 7836   14._____

Name: Date:
Instructor: Section:

Objective C Write the word name for a standard numeral.

Write the word name for each of the following standard numerals.

15. 18   15._____

16. 347   16._____

17. 398,410   17._____

18. 20,001   18._____

19. 562   19._____

20. 65,805,300   20._____

Write each of the following as a standard numeral.

21. Six thousand, four hundred eighty-seven   21._____

22. Fifty-three million, seven   22._____

Copyright © 2013 Pearson Education, Inc.

23. Eight hundred forty 23._____

24. Ten thousand, six hundred three 24._____

25. One hundred eighty-two million, seven hundred 25._____
four thousand, nine

Name: Date:
Instructor: Section:

Chapter R   BASIC IDEAS

R.2   Addition and Subtraction of Whole Numbers

Learning Objectives
A   Find the sum of two digits from memory.
B   Add whole numbers.
C   Use combinations of numbers to add columns of whole numbers.
D   Find the difference of two digits from memory.
E   Subtract whole numbers without regrouping (borrowing).
F   Subtract whole numbers with regrouping (borrowing).
G   Solve application problems using subtraction.

Getting Ready

Construct standard numerals for each of the following whole numbers from word notation.

1. Six thousand, four hundred twenty-two   1._____

2. Seven hundred eighty   2._____

3. Two million, nine hundred nineteen thousand,   3._____
   forty-one

Key Terms

Use the vocabulary terms listed below to complete each statement in exercises 1–9.

| addends | minuend | subtraction | sum | zero |
| subtrahend | variable | difference | addition | |

1. A(n) _____ is a symbol, usually a letter, that stands for the numbers of a specified set.

2. _____ subtracted from a number yields the original number.

3. The concept of _____ may be thought of as putting two nonoverlapping sets together to form a combined set.

4. _____ may be thought of as a taking away of objects from a set.

5. In the sentence $a+b=c$, the numbers a and b are called _____.

Copyright © 2013 Pearson Education, Inc.   5

6. In the sentence $a+b=c$, the number c is called the _____.

7. In the sentence $a-b=c$, the number a is called the _____.

8. In the sentence $a-b=c$, the number b is called the _____.

9. In the sentence $a-b=c$, the number c is called the _____.

Practice Problems

Objectives A, B, C Find the sum of two digits from memory. Add whole numbers. Use combinations of numbers to add columns of whole numbers.

Add each of the following.

1. 632
   +51

   1._____

2. 2837
   +162

   2._____

3. 45
   +32

   3._____

4. 1072
   + 25

   4._____

Add each of the following.

5. 769
   +847

   5._____

6. 1658
   +2993

   6._____

Name: Date:
Instructor: Section:

7.   17,612
  + 4,854

7._____

8.   87,099
  + 6,147

8._____

9.   156
    38
 +297

9._____

10.   39
   58
   46
 +17

10._____

11.   127
   34
  781
 + 65

11._____

12.   7
  1086
   434
    58
 + 9625

12._____

Objectives D, E, F **Find the difference of two digits from memory. Subtract whole numbers without regrouping (borrowing). Subtract whole numbers with regrouping (borrowing).**

Subtract. Check each difference with addition.

13. 896 − 53

13._____

Copyright © 2013 Pearson Education, Inc.

14. $7846 - 435$ 14._____

15. $76,589 - 1476$ 15._____

16. $7653 - 5412$ 16._____

17. $83 - 59$ 17._____

18. $164 - 78$ 18._____

19. $1007 - 984$ 19._____

20. $4311 - 299$ 20._____

21. $8059 - 7972$ 21._____

22. $312 - 268$ 22._____

Objective G   Solve application problems using subtraction.

Solve the following word problems.

23. Patty's math class has 28 students. 17 of the students are male. How many of the students are female?   23._____

24. Jacob has $1489 in his savings account. He deposited a $275 bonus he received from his job into that savings account. What is the new balance in Jacob's account?   24._____

Name:
Instructor:
Date:
Section:

Chapter R BASIC IDEAS

R.3 Multiplication and Division of Whole Numbers

Learning Objectives
A Find the product of two numbers.
B Know the multiplication facts through 12 by 12.
C Use an algorithm to multiply in standard notation.
D Recognize and apply the relationship between multiplication and division.
E Recognize that division by zero is not possible.
F Divide whole numbers.

Getting Ready

Add or subtract each of the following whole numbers:

1. 409
   +572

   1._____

2. $64 + 64 + 64 + 64$

   2._____

3. 73
   −16

   3._____

4. 654
   −318

   4._____

Key Terms

Use the vocabulary terms listed below to complete each statement in exercises 1–6.

| product | factors | multiples | quotient | divisor | dividend |

1. In a multiplication statement the numbers being multiplied are called the _____.

2. In the statement $a \div b = c$, the number a is called the _____.

3. The _____ of a number are the results of multiplying the number by each of the whole numbers, starting with 0.

4. In the statement $a \div b = c$, the number b is called the _____.

5. In the sentence $a \cdot b = c$, the number c is called the _____.

6. In the statement $a \div b = c$, the number c is called the _____.

Practice Problems

Objectives A, B, C Find the product of two numbers. Know the multiplication facts through 12 by 12. Use an algorithm to multiply in standard notation.

Multiply.

1. 12(38)
1._____

2. 7(59)
2._____

3. 9(137)
3._____

4. 67
 × 15
4._____

5. 86
 × 54
5._____

6. 546
 × 20
6._____

Name:
Instructor:

Date:
Section:

7. 509
   × 32

7._____

8. 607
   × 55

8._____

9. 183
  × 704

9._____

10. 418
   × 703

10._____

Solve each problem.

11. In a typical bowling center there are 22 bowling pins for use behind each lane. Airport Bowl has 32 lanes. They want to order new pins to replace all their used ones. How many bowling pins will they need to order?

11._____

12. South University has a building with 24 classrooms. Each classroom contains 8 rows of 6 desks. How many desks are in this particular building?

12._____

Objectives D, E, F Recognize and apply the relationship between multiplication and division. Recognize that division by zero is not possible. Divide whole numbers.

Divide by using multiples of the divisor and check by using related multiplication.

13. $63 \div 5$ 13._____

14. $94 \div 13$ 14._____

15. $47 \div 9$ 15._____

16. $56 \div 17$ 16._____

Divide and check by using the related multiplication.

17. $5\overline{)427}$ 17._____

18. $8\overline{)961}$ 18._____

Name:
Instructor:

Date:
Section:

19. $11\overline{)126}$

19. _____

20. $40\overline{)268}$

20. _____

21. $512\overline{)5466}$

21. _____

22. $0\overline{)78}$

22. _____

Solve.

23. A shuttle bus must transport 224 people from a parking lot to a conference center. The bus holds 32 people. How many trips will it take to transport all of the people?

23. _____

24. A family of four won a $3748 prize in a contest. They decided to split the money evenly amongst themselves. How much money did each family member receive?

24._____

25. Paul and three of his friends went out for dinner. They decided to split the $112 check evenly amongst themselves. How much did each person pay?

25._____

Name: Date:
Instructor: Section:

Chapter R BASIC IDEAS

R.4 A Brief Introduction to Fractions

Learning Objectives
A Write a fraction that represents a given part of a whole (unit).
B Identify the part of the whole (unit) that is represented by a fraction.
C Change an improper fraction into a mixed number and a mixed number into an improper fraction.
D Reduce a fraction to lowest terms.
E Multiply two fractions.
F Divide two fractions.
G Add and subtract fractions with the same denominators.

Getting Ready

Add or subtract the following whole numbers:

1. $7 + 4 + 6$ 1._____

2. $63 - 38$ 2._____

Multiply or divide the following whole numbers:

3. $(8)(7)$ 3._____

4. $28 \div 4$ 4._____

5. Match one of the terms *sum, difference, product,* and   5a._____
   quotient with the results of each of the following
   operations.   b._____
   a. $7 - 4$ b. $(3)(5)$
                                                               c._____
   c. $6 \div 3$ d. $2 + 8$
                                                               d._____

Copyright © 2013 Pearson Education, Inc.   15

Key Terms

Use the vocabulary terms listed below to complete each statement in exercises 1–6.

    lowest terms   **divide**   **multiplicative inverses**

    numerator   **denominator**   **fraction**

1. A _____ is a number in the form $\frac{a}{b}$ with $b \neq 0$.

2. In the number $\frac{a}{b}$, a is called the _____.

3. In the number $\frac{a}{b}$, b is called the _____.

4. Two numbers are _____ if their product is 1.

5. A fraction is in _____ if the only whole number that will divide the numerator and the denominator is 1.

6. To _____ by a fraction, multiply by its reciprocal.

Practice Problems

Objective A Write a fraction that represents a given part of a whole (unit).

Using a fraction, represent the part of the whole region that is shaded.

1.

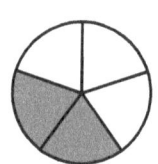

    1._____

2.

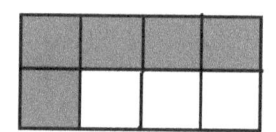

    2._____

Name:
Instructor:

Date:
Section:

Objective B Identify the part of the whole (unit) that is represented by a fraction.

Shade the part of the whole that is represented by the given fraction.

3. $\dfrac{7}{8}$

3._____

4. $\dfrac{2}{6}$

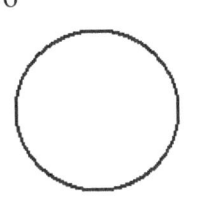

4._____

Objective C Change an improper fraction into a mixed number and a mixed number into an improper fraction.

Change the improper fractions into mixed numbers.

5. $\dfrac{41}{9}$

5._____

6. $\dfrac{51}{8}$

6._____

7. $\dfrac{37}{7}$

7._____

8. $\dfrac{116}{15}$

8._____

Change the mixed numbers into improper fractions.

9. $2\dfrac{3}{5}$

9._____

10. $11\frac{4}{7}$

10._____

11. $14\frac{5}{6}$

11._____

12. $21\frac{4}{9}$

12._____

Objective D Reduce a fraction to lowest terms.

Reduce to lowest terms.

13. $\dfrac{12}{42}$

13._____

14. $\dfrac{18}{76}$

14._____

15. $\dfrac{35}{50}$

15._____

16. $\dfrac{24}{96}$

16._____

Objective E Multiply two fractions.

Find the following products.

17. $\dfrac{1}{2} \cdot \dfrac{1}{9}$

17._____

18. $\dfrac{2}{7} \cdot \dfrac{3}{11}$

18._____

19. $\dfrac{2}{17} \cdot \dfrac{3}{5}$

19._____

20. $\dfrac{8}{15} \cdot \dfrac{2}{3}$

20._____

Name:
Instructor:

Date:
Section:

21. $8 \cdot \dfrac{1}{4}$

21._____

22. $\dfrac{1}{7} \cdot 35$

22._____

Objective F Divide two fractions.

Find the following quotients.

23. $9 \div \dfrac{1}{4}$

23._____

24. $18 \div \dfrac{1}{3}$

24._____

25. $\dfrac{2}{3} \div 11$

25._____

26. $\dfrac{5}{6} \div 13$

26._____

27. $\dfrac{7}{8} \div \dfrac{5}{3}$

27._____

28. $\dfrac{9}{16} \div \dfrac{4}{3}$

28._____

29. $3 \div \dfrac{2}{7}$

29._____

30. $\dfrac{8}{9} \div \dfrac{15}{4}$

30._____

Objective G Add and subtract fractions with the same denominators.

Find the following sums and differences.

31. $\dfrac{1}{9} + \dfrac{4}{9}$

31._____

Copyright © 2013 Pearson Education, Inc.

32. $\dfrac{2}{13}+\dfrac{5}{13}$

32._____

33. $\dfrac{6}{7}-\dfrac{1}{7}$

33._____

34. $\dfrac{7}{11}-\dfrac{4}{11}$

34._____

35. $\dfrac{5}{21}+\dfrac{8}{21}$

35._____

36. $\dfrac{13}{31}-\dfrac{2}{31}$

36._____

37. $\dfrac{1}{14}+\dfrac{5}{14}-\dfrac{3}{14}$

37._____

38. $\dfrac{14}{51}+\dfrac{23}{51}+\dfrac{1}{51}$

38._____

Answer the following.

39. A recipe for a batch of cookies calls for $\dfrac{3}{4}$ cup of chocolate chips. Find the number of cups of chocolate chips needed for 7 such batches.

39._____

40. A certain stock rose $\dfrac{3}{8}$ point on Monday, and $\dfrac{5}{8}$ point on Tuesday, then fell $\dfrac{1}{8}$ point on Wednesday. What was the point change in this stock over the three days?

40._____

Name: Date:
Instructor: Section:

Chapter R   BASIC IDEAS

R.5   Addition and Subtraction of Decimal Numerals

Learning Objectives
A   Recognize, read, and write decimals in standard notation.
B   Determine which decimal is larger or largest.
C   Add decimals.
D   Subtract decimals.

Getting Ready

Add or subtract the following whole numbers, as indicated:

1. $618 + 475$ 1._____

2. $1956 - 482$ 2._____

3.   702   3._____
      49
   +1513

Multiply or divide the following numbers as indicated:

4. $\dfrac{1}{10} \cdot \dfrac{1}{1000}$ 4._____

5. $\dfrac{10{,}000}{100}$ 5._____

6. Give the names of the results in exercises 1–5: sum, difference, product, or quotient.   6._____

7. It takes Joy 45 minutes to ride her bike from her home to school. She has been riding for 20 minutes. How much longer must she ride until she gets to school?   7._____

Copyright © 2013 Pearson Education, Inc.

Key Terms

Use the vocabulary terms listed below to complete each statement in exercises 1–3.

    place values   **decimal point**   **decimal mixed numerals**

1. Decimal numerals that include whole numbers and decimal fractions are called _____.

2. When adding and subtracting decimals line up corresponding _____.

3. The first place value to the right of the _____ is the tenths place.

Practice Problems

Objective A Recognize, read, and write decimals in standard notation.

For each of the following write the word name.

1. .51   1._____

2. 1.695   2._____

3. 78.305   3._____

Identify the place value of the underlined digit.

4. 2.90<u>3</u>   4._____

5. 56.<u>1</u>23   5._____

6. 326.0152<u>6</u>   6._____

Objective B Determine which decimal is larger or largest.

Decide which decimal numeral in each exercise represents the larger or the largest number.

7. .316, .328   7._____

Name: Date:
Instructor: Section:

8. 16.5432, 16.548

8._____

9. 2.0013, 2.0107, 2.06009

9._____

10. 42,308, 4397, 465.9998

10._____

Objective C Add decimals.

Add each of the following.

11. .3 + .9

11._____

12. .71 + .68

12._____

13. 2.142 + 7.49

13._____

14. 12.9 + 13.82

14._____

15. 625.63 + 408.371

15._____

16. 19.27 + 158.065

16._____

17. 4.16 + 1.078 + .0449

17._____

18. 71.06 + 8.543 + 19.112 + 26.379

18._____

Copyright © 2013 Pearson Education, Inc.

Objective D Subtract decimals.

Subtract each of the following.

19. .6 − .13

19._____

20. .514 − .3816

20._____

21. 2.731 − 1.06

21._____

22. 39 − .0017

22._____

23. 261.0844 − 47

23._____

24. 14.208 − 8.0904

24._____

Write the word name for each of the following money amounts as it would appear on a check.

25. $247.38

25._____

26. $80.54

26._____

Solve each of the following.

27. Philip wrote three checks amounting to $5.41, $161.76, and $27.98. What is the total amount of money he must deduct from his check register due to these three checks?

27._____

28. Marco had two pieces of rope. One piece was 6.9 feet long and the other measured 14.3 feet. What was the combined length of Marco's ropes?

28._____

Name: Date:
Instructor: Section:

Chapter R BASIC IDEAS

R.6 Multiplication and Division of Decimal Numerals

Learning Objectives
A Multiply decimals.
B Divide decimals.
C Round decimals.
D Divide decimals and round the quotient.
E Represent a fraction as division and rewrite a fraction as a decimal.

Getting Ready

Multiply each of the following whole numbers:

1. $\begin{array}{r} 36 \\ \times\ 17 \end{array}$ 1._____

2. $\begin{array}{r} 529 \\ \times\ \ \ 6 \end{array}$ 2._____

Divide each of the following whole numbers:

3. $108 \div 6$ 3._____

4. $228 \div 12$ 4._____

Find the product.

5. $\dfrac{1}{10,000} \cdot \dfrac{1}{100}$ 5._____

Key Terms

Use the vocabulary terms listed below to complete each statement in exercises 1–3.

    whole number   **rounding**   **sum**

1. The _____ of the number of decimal places in all of the factors is the number of decimal places in the product.

2. When dividing by a decimal start by making the divisor a _____.

3. The process of _____ involves writing a numeral to the nearest desired place value.

Practice Problems

Objective A   Multiply decimals.

Multiply.

1. 5(2.9)

    1._____

2. .71(.8)

    2._____

3. .4(.6)

    3._____

4. .19(7.8)

    4._____

5. 21.6(.123)

    5._____

6. 6(.45)

    6._____

7. 16(.015)

    7._____

8. 3.8(.003)

    8._____

Name: Date:
Instructor: Section:

Objective B Divide decimals.

Divide.

9. $4.9 \div .7$　　　　　　　　　　　　　　　　　　9._____

10. $.64 \div .008$　　　　　　　　　　　　　　　　10._____

11. $1.21 \div 1.1$　　　　　　　　　　　　　　　　11._____

12. $3.84 \div .3$　　　　　　　　　　　　　　　　　12._____

Objective C Round decimals.

Round to the indicated place value.

13. .637   nearest hundredth　　　　　　　　　　13._____

14. 28.0461   nearest tenth　　　　　　　　　　14._____

15. 1.0008   nearest thousandth　　　　　　　　15._____

16. 14.58   nearest one　　　　　　　　　　　　16._____

Objective D Divide decimals and round the quotient.

Divide. If a quotient goes beyond the hundredths place, round to the nearest hundredth.

17. $12.84 \div 7$ 17._____

18. $3.061 \div .16$ 18._____

19. $10.005 \div .15$ 19._____

20. $.623 \div 4.02$ 20._____

Objective E Represent a fraction as division and rewrite a fraction as a decimal.

Change the fractions into decimals. If necessary, round answers to the nearest hundredth.

21. $\dfrac{1}{5}$ 21._____

22. $\dfrac{1}{8}$ 22._____

23. $\dfrac{5}{12}$ 23._____

Name:
Instructor:
Date:
Section:

24. $\dfrac{7}{15}$

24._____

25. $\dfrac{14}{25}$

25._____

26. $\dfrac{56}{48}$

26._____

Solve. Round to the nearest hundredth where necessary.

27. $1433.50 was collected from 47 people. If they each donated the same amount, how much did each donate?

27._____

28. Frankie has seven days to complete 230 math problems. What is the average number of problems he needs to complete each day?

28._____

29. West Theatre sold 246 tickets. 120 tickets sold for $22 each while the remaining 126 tickets sold for $16.50 each. How much money was collected from the sale of tickets?

29._____

30. Kenny loaded 18 boxes each weighing 108.6 pounds onto a flatbed trailer. What was the total weight of the boxes?

30. _____

Name: Date:
Instructor: Section:

Chapter R   BASIC IDEAS

R.7   Linear Measurement in the American and Metric Systems

Learning Objectives
A   Convert linear units within the American system of measurement.
B   Convert linear units within the metric system of measurement.

Getting Ready

Perform each indicated operation.

1. $4.8 \cdot 12$ 1._____

2. $5.19 \cdot 1000$ 2._____

3. $6.17 \div 100$ 3._____

4. $8296 \div 1000$ 4._____

Multiply the following:

5. $10 \cdot \dfrac{12}{1}$ 5._____

6. $648 \cdot \dfrac{1}{36}$ 6._____

7. $15,840 \cdot \dfrac{1}{5280}$ 7._____

8. $7.8 \cdot \dfrac{36}{1}$ 8._____

Copyright © 2013 Pearson Education, Inc.

Key Terms

Use the vocabulary terms listed below to complete each statement in exercises 1–4.

American   **metric**   **unit-cancellation**   **meter**

1. The basic unit of linear measure in the metric system is the _____.

2. One of the greatest advantages of the _____ system is that conversions from one linear unit of measure to another are accomplished by powers of 10.

3. One method of converting linear units within a system of measurement is called the _____ method.

4. In the _____ system of linear measure, some of the units are inch, foot, and yard.

Practice Problems

Objective A Convert linear units within the American system of measurement.

Convert the following to the indicated unit.

1. 5 ft = _____ in.   1._____

2. 21,120 ft = _____ mi   2._____

3. 23 yd = _____ ft   3._____

4. 18 mi = _____ ft   4._____

5. 8.5 mi = _____ yd   5._____

Name: Date:
Instructor: Section:

6. 4.8 yd = _____ in. 6._____

7. 1476 in. = _____ yd 7._____

8. 720 in. = _____ ft 8._____

Objective B Convert linear units within the metric system of measurement.

Convert the following to the indicated unit.

9. 260 mm = _____ cm 9._____

10. 6470 cm = _____ km 10._____

11. 2.618 km = _____ dm 11._____

12. 43.7 hm = _____ mm 12._____

13. 930 dam = _____ dm 13._____

14. .00006 km = _____ mm 14._____

15. 14 cm = _____ m   **15.** _____

16. 8 dm = _____ dam   **16.** _____

Name: Date:
Instructor: Section:

Chapter 1 ADDING AND SUBTRACTING INTEGERS AND POLYNOMIALS

1.1 Variables, Exponents, and Order of Operations

Learning Objectives
A   Identify variables and constants.
B   Read and evaluate expressions raised to powers.
C   Simplify expressions involving more than one operation.
D   Evaluate expressions containing variables, when given the value(s) of the variable(s).

Getting Ready

Add, subtract, multiply, or divide the following whole numbers:

1. $17 + 56$
   1._____

2. $72 - 46$
   2._____

3. $8 \cdot 8 \cdot 8$
   3._____

4. $\dfrac{80}{16}$
   4._____

Key Terms

Use the vocabulary terms listed below to complete each statement in exercises 1–5.

    exponent   base   π   constant   variable

1. _____ is a constant whose approximate value is 3.14.

2. A(n) _____ is a symbol, usually a letter, that is used to represent a number.

3. In the expression 2^5, 2 is called the _____ .

4. In the expression 2^5, 5 is called the _____ .

5. A(n) _____ is a symbol (a numeral) whose value does not change.

Practice Problems

Objective A Identify variables and constants.

Identify the constant and the variable(s) in each of the following.

1. $9x$ 1._____

2. $2ab$ 2._____

3. t 3._____

4. 15 4._____

Objective B Read and evaluate expressions raised to powers.

Write a statement indicating how each of the following would be read, and then evaluate.

5. 7^3 5._____

6. 9^2 6._____

Write each of the following as multiplication.

7. x^6 7._____

8. $2y^4$ 8._____

9. $3a^2b^5$ 9._____

10. $(9p)^3$ 10._____

Write each of the following using exponents.

11. $x \cdot x \cdot x$ 11._____

12. $p \cdot p \cdot q \cdot q \cdot q \cdot q$ 12._____

13. $7c \cdot c \cdot c \cdot d \cdot d$ 13._____

Name: Date:
Instructor: Section:

14. $(8m)(8m)(8m)(8m)(8m)$ 14._____

Objective C Simplify expressions involving more than one operation.

Simplify the following using the order of operations.

15. $9-6+4$ 15._____

16. $5-12\div 6$ 16._____

17. $3+2\cdot 9$ 17._____

18. $17+(19-5)\div 7$ 18._____

19. $3^2\cdot 4^3$ 19._____

20. $48\div 6\cdot 5+12$ 20._____

21. $40\div 10\cdot 9-29$ 21._____

22. $6(3^4-30)+21$ 22._____

23. $5(6^2)-18$ 23._____

24. $8 + 42 \div 6 \cdot 3 - 7$

24._____

25. $\dfrac{7 + 3 \cdot 4}{19}$

25._____

26. $\dfrac{4^2 + 3^2 - 7}{81 \div 9}$

26._____

Objective D Evaluate expressions containing variables, when given the value(s) of the variables(s).

Let x = 4 and y = 5 and evaluate.

27. $12x - 3y$

27._____

28. $5x^2$

28._____

29. $8xy^2$

29._____

30. $\dfrac{2x + 7y + 1}{x^2 - y}$

30._____

Write an expression for each of the following and evaluate.

31. Roy bought a freezer by paying $240 down and $52 per month for 12 months. Find the cost of the freezer.

31._____

32. Martin and his four bowling partners agree to split the cost equally for two pizzas that cost $12.50 each, 4 orders of breadsticks that cost $2.75 each, and five sodas that cost $1.25 each. Find the amount (to the nearest cent) that each pays.

32._____

Name:
Instructor:
Date:
Section:

Chapter 1 ADDING AND SUBTRACTING INTEGERS AND POLYNOMIALS

1.2 Perimeters of Geometric Figures

Learning Objectives
A   Find the perimeters of geometric figures including triangles, squares, and rectangles, and the circumference of a circle.
B   Solve application problems involving perimeters of geometric figures.

Getting Ready

Add, multiply, or divide the following whole numbers:

1. $16 + 34 + 25$ 1._____

2. $4(3.14)(6)$ 2._____

3. $8(10) + 8(12)$ 3._____

4. $\dfrac{39 \cdot 11}{13}$ 4._____

Substitute the given value(s) of the variable(s) and evaluate by using the order of operations.

5. $2\pi r$: $\pi \approx \dfrac{22}{7}$, $r = 49$ 5._____

6. $2l + 2w$: $l = 8$, $w = 9$ 6._____

Key Terms

Use the vocabulary terms listed below to complete each statement in exercises 1–4.

    diameter   π   perimeter   circumference

1. The _____ of a geometric figure is the distance around it.

2. The distance around a circle is called the _____.

Copyright © 2013 Pearson Education, Inc.   39

3. A line segment with endpoints on the circle and passing through the center is called a _____.

4. The ratio of the circumference to the diameter of any circle is_____.

Practice Problems

Objective A   Find the perimeters of geometric figures including triangles, squares, and rectangles, and the circumference of a circle.

Find the perimeter of each of the following shapes.

1.

   2 in.
1 in.   3 in.
2 in.
           3 in.
   4 in.

1._____

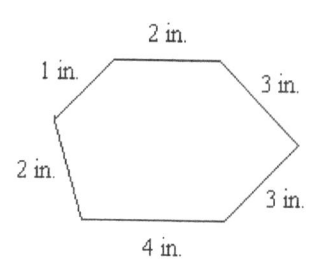

2.

   7 ft
3 ft
       2 ft
   10 ft

2._____

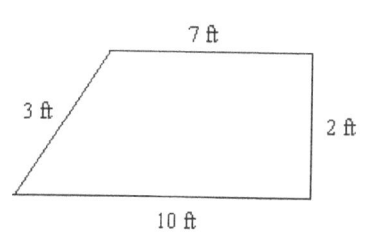

Find the perimeter of each of the following rectangles.

3.

   9 m
4 m

3._____

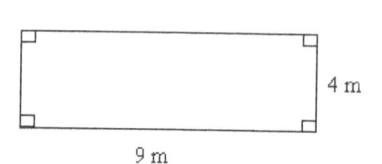

4.

6 ft
9 ft

4._____

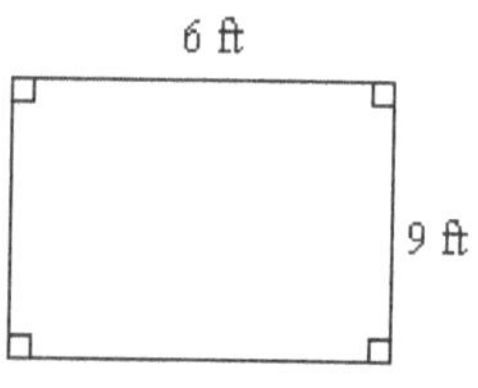

Name: Date:
Instructor: Section:

5. Length of 12 yards and width of 15 feet. Express the answer in yards.

5._____

6. Length of 3 feet and width of 26 inches. Express the answer in inches.

6._____

Find the perimeter of each of the following squares.

7.

7._____

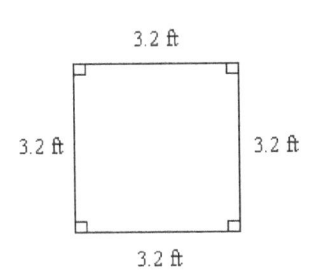

3.2 ft
3.2 ft   3.2 ft
3.2 ft

8. Each of whose sides is 19 kilometers long.

8._____

Find the perimeter of the following triangles.

9.

9._____

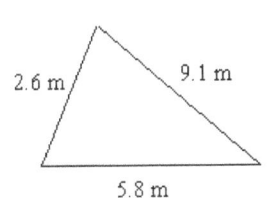

2.6 m   9.1 m
5.8 m

10.

10._____

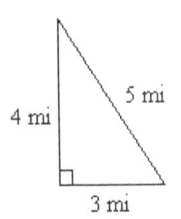

5 mi
4 mi
3 mi

11. A triangle whose sides are 16 feet, 12 feet, and 24 feet in length.

11._____

Find the exact value (express answers in terms of π) and the approximate value (use π ≈ 3.14) of the circumference of each of the following circles.

12.

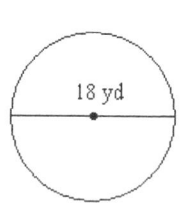

    4 dm

12. _____

13.

    18 yd

13. _____

14. A circle whose radius is 3.5 feet.

14. _____

15. A circle with a diameter of 34 inches.

15. _____

Objective B Solve application problems involving perimeters of geometric figures.

Solve each problem. If necessary, use π ≈ 3.14 and round your answers to the nearest hundredth of a unit.

16. John wants to build a sandbox. The rectangular dimensions are 6 feet by 8 feet. How much material does he need to make the frame?

16. _____

17. Meghan has a clock whose diameter is 6 inches. What is the circumference of Meghan's clock?

17. _____

18. Andrea wants to put a wood rail border around her flower garden. The garden is rectangular and has dimensions 3 feet by 16 feet. How much wood railing should she purchase?

18. _____

19. A truck tire has a diameter of 32 inches. How far will the tire roll in eight revolutions?

19. _____

20. Carol wants to put a ribbon around the edge of a circular wreath. The wreath has a diameter of 16 inches. The ribbon costs $2.45 per foot. How much will she need to spend on ribbon?

20. _____

Name:
Instructor:

Date:
Section:

Chapter 1 ADDING AND SUBTRACTING INTEGERS AND POLYNOMIALS

1.3 Areas of Geometric Figures

Learning Objectives
A   Find the areas of rectangles, squares, triangles, parallelograms, trapezoids, and circles.
B   Solve application problems involving area.

Getting Ready

Multiply or divide the following whole numbers and/or decimals.

1. $(17)(3)$ 1._____

2. $(15)(3.14)$ 2._____

3. $\dfrac{(6.9)(5)}{3}$ 3._____

Substitute the given value(s) of the variable(s) and evaluate by using the order of operations.

4. $\dfrac{bh}{2}$: $b = 4.2$, $h = 7.1$ 4._____

5. πr^2: $\pi \approx 3.14$, $r = 2$ 5._____

6. $\dfrac{h(B+b)}{2}$: $h = 9$, $B = 15$, $b = 3$ 6._____

Key Terms

Use the vocabulary terms listed below to complete each statement in exercises 1–8.

| square unit | right | area | $A = s^2$ |
| $A = \dfrac{h(B+b)}{2}$ | $A = LW$ | $A = \dfrac{bh}{2}$ | $A = \pi r^2$ |

1. The measure of the size of a surface is called its_____.

2. A(n) _____ is a square each of whose sides is one linear unit.

Copyright © 2013 Pearson Education, Inc.   43

3. An angle that measures 90° is called a(n) _____ angle.

4. The formula for area of a triangle is _____.

5. The formula for area of a square is _____.

6. The formula for area of a trapezoid is _____.

7. The formula for area of a rectangle is _____.

8. The formula for area of a circle is _____.

Practice Problems

Objective A Find the areas of rectangles, squares, triangles, parallelograms, trapezoids, and circles.

Find the area of each of the following rectangles.

1.

    4 m

    9 m

1._____

2.

    6 ft

    9 ft

2._____

3. Length of 12 yards and width of 15 feet. Express the answer in square yards.

3._____

4. Length of 3 feet and width of 26 inches. Express the answer in square inches.

4._____

Name:
Instructor:

Date:
Section:

Find the area of each of the following squares.

5.

    3.2 ft
3.2 ft   3.2 ft
    3.2 ft

5. _____

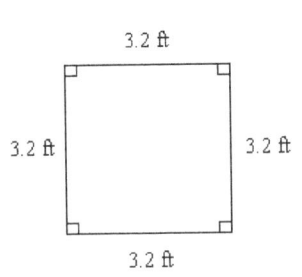

6. Each side is 19 kilometers long.

6. _____

Find the area of the following triangles.

7.

8 ft, 14 ft, 5 ft, 19 ft

7. _____

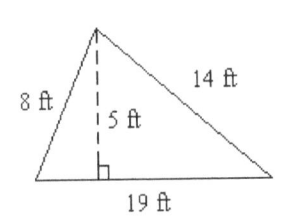

8.

4 mi, 5 mi, 3 mi

8. _____

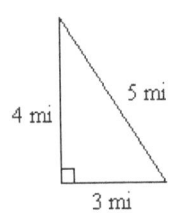

9. A triangle with two sides that are each 20 feet in length and a third side 24 feet in length, and whose height drawn to the 24-foot side is 16 feet.

9. _____

Copyright © 2013 Pearson Education, Inc.

45

Find the exact value (express answers in terms of π) and the approximate value (use π ≈ 3.14) of the area of the following circles. Round approximate answers to the nearest tenth of a unit.

10.

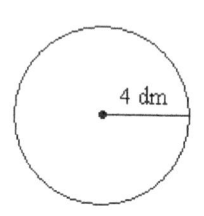

4 dm

10._____

11.

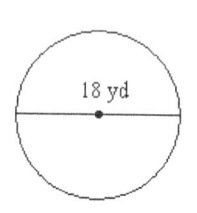

18 yd

11._____

12. A circle whose radius is 6.6 feet.

12._____

13. A circle with a diameter of 34 inches.

13._____

Find the area of each of the following parallelograms.

14.

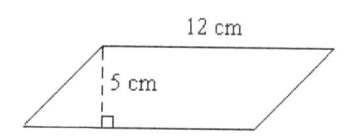

12 cm
5 cm

14._____

15. A parallelogram whose base is 19 feet and whose height is 72 inches. Express the answer in square feet.

15._____

Name:	Date:
Instructor:	Section:

Find the area of each of the following trapezoids.

16.

12 cm
24 cm
22 cm
30 cm

16._____

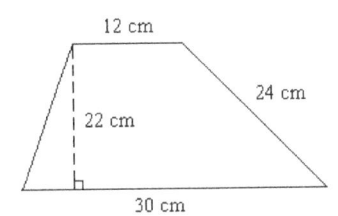

17.

12.3 ft
4 ft
6.2 ft

17._____

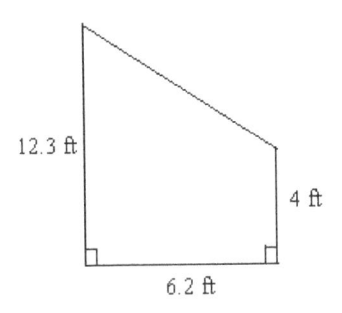

Objective B Solve application problems involving area.

Solve.

18. A math classroom is 20 feet long and 14 feet wide. If carpet costs $22.30 per square yard, find the cost of the carpet needed to carpet the floor of the classroom.

18._____

19. Betsy wants to make a mat to go under her table cloth for her round dining table. If her table has a diameter of 6 feet, how many square feet of mat will she have?

19._____

20. Jay's Hardware sells a gallon of water seal for $18. One gallon of seal will cover 200 square feet. Jack wants to put one coat of water seal on his deck that measures 32 feet by 50 feet. How much money will he have to spend on water seal?

20._____

Name: Date:
Instructor: Section:

Chapter 1 ADDING AND SUBTRACTING INTEGERS AND POLYNOMIALS

1.4 Volumes and Surface Areas of Geometric Figures

Learning Objectives
A   Find the volumes of rectangular solids, cubes, right circular cylinders, and cones.
B   Find the surface area of rectangular solids, cubes, and right circular cylinders.
C   Solve application problems involving volumes and surface areas.

Getting Ready

Add, multiply, or divide the following whole numbers and/or decimals:

1. $(9)(13)(6)$ 1._____

2. $\dfrac{5(3.14)(12)}{3}$ 2._____

Substitute the given value(s) of the variable(s) and evaluate using the order of operations.

3. $\dfrac{\pi r^2 h}{3}$: $\pi \approx 3.14$, $r = 6$, $h = 8$ 3._____

4. $\dfrac{4}{3}\pi r^3$: $\pi \approx 3.14$, $r = 6$ 4._____

Key Terms

Use the vocabulary terms listed below to complete each statement in exercises 1–8.

surface area	volume	sphere	cube
$V = LWH$	$V = \pi r^2 h$	$V = e^3$	$V = \dfrac{4}{3}\pi r^3$

1. When finding the _____ of a solid, we are finding the number of cubic units it takes to fill the solid.

2. A _____ is a three-dimensional geometric solid with six faces, each of which is a square.

3. The formula for volume of a right circular cylinder is _____.

4. The formula for volume of a rectangular solid is _____.

5. The set of all points in space that are equidistant from a given point is called a _____.

6. The formula for volume of a cube is _____.

7. The formula for volume of a sphere is _____.

8. $2LW + 2LH + 2WH$ is used to find the _____ of a rectangular solid.

Practice Problems

Objectives A, B Find the volumes of rectangular solids, cubes, right circular cylinders, and cones. Find the surface area of rectangular solids, cubes, and right circular cylinders.

Find the volume and surface area of each of the following. Round answers to the nearest hundredth.

1.

6 in.
4 in.
9 in.

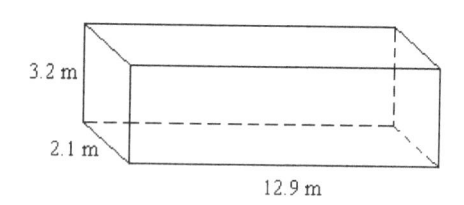

1._____

2.

3.2 m
2.1 m
12.9 m

2._____

Name:
Instructor:

Date:
Section:

3.

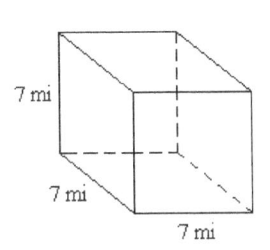

7 mi
7 mi
7 mi

3._____

4.

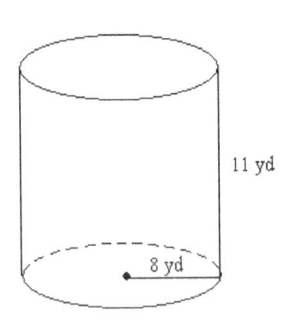

11 yd
8 yd

4._____

5.

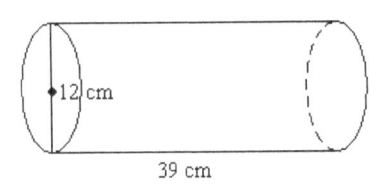

12 cm
39 cm

5._____

6. A cube each of whose edges is $\frac{2}{3}$ foot.

6._____

7. A right circular cylinder whose base radius is 6 centimeters and whose height is 14 centimeters.

7._____

Find the volume of each of the following. Round answers to the nearest tenth.

8.

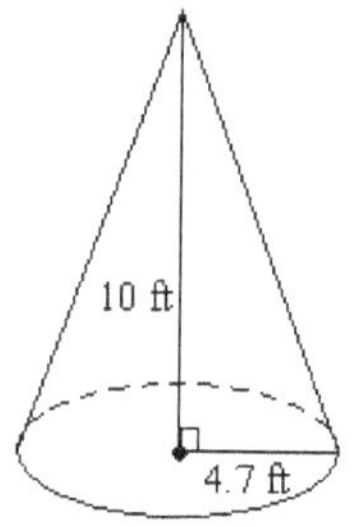

10 ft

4.7 ft

8._____

9.

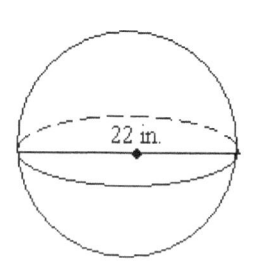

22 in.

9._____

10. A sphere whose diameter is 18 inches.

10._____

Name:
Instructor:

Date:
Section:

Objective C Solve application problems involving volumes and surface areas.

Solve.

11. A storage trunk is 1.5 feet wide, 2 feet high, and 4 feet long. How many cubic feet will the trunk hold?

    11._____

12. A propane gas tank is in the shape of a right circular cylinder. It is 8 feet long with a diameter of 4 feet. What is the exact volume of the tank?

    12._____

13. Chad wants to paint a rectangular box that is 2 feet by 4 feet by 8 feet. How many square feet will he paint?

    13._____

Copyright © 2013 Pearson Education, Inc.

Name: Date:
Instructor: Section:

Chapter 1   ADDING AND SUBTRACTING INTEGERS AND POLYNOMIALS

1.5   Introduction to Integers

Learning Objectives
A   Identify natural numbers, whole numbers, and integers.
B   Graph natural numbers, whole numbers, and integers.
C   Determine order relations for integers.
D   Find the negative (opposite) of an integer.
E   Find the absolute value of an integer.

Key Terms

Use the vocabulary terms listed below to complete each statement in exercises 1–6.

     absolute value   natural   opposites
     whole   graphing   integers

1. $\{1, 2, 3, 4, 5, ...\}$ is the set of _____ numbers.

2. $\{0, 1, 2, 3, 4, 5, ...\}$ is the set of _____ numbers.

3. $\{..., -3, -2, -1, 0, 1, 2, 3, ...\}$ is the set of _____.

4. -5 and 5 are _____ of each other.

5. When an integer is paired with a point on the number line, this procedure is called _____.

6. The _____ of a number is defined as the distance from 0 to the number on the number line.

Practice Problems

Objective A   Identify natural numbers, whole numbers, and integers.

Answer each of the following using signed numbers.

1. If 75 miles represents +75 miles north, what would   1._____
   −75 miles represent?

2. If −10 points represents losing 10 points, what would   2._____
   +10 points represent?

Objective B Graph natural numbers, whole numbers, and integers.

Draw a number line for each of the following and graph the indicated integers.

3. {−3, −1, 0, 2}   3. See graph

    ←+++++++++++++++++++++++++++→

4. {−6, −4, 1, 5}   4. See graph

    ←+++++++++++++++++++++++++++→

Objective C Determine order relations for integers.

Insert the proper symbol =, <, or > in order to make each of the following true.

5. 5 ____ 9   5. _____

6. 18 ____ 10   6. _____

7. −6 ____ −4   7. _____

8. 14 ____ −5   8. _____

9. $|-2|$ ____ 2   9. _____

10. $|-4|$ ____ −4   10. _____

11. $|15-7|$ ____ $|10-3|$ 11. _____

12. −(−15) ____ 15   12. _____

56   Copyright © 2013 Pearson Education, Inc.

Name: Date:
Instructor: Section:

Objective D Find the negative (opposite) of an integer.

Find the negative (opposite) of each of the following.

13. 21

13._____

14. −19

14._____

15. $|8|$

15._____

16. $|-21|$

16._____

17. $-|7|$

17._____

Objective E Find the absolute value of an integer.

Evaluate the following.

18. $|16-21|$

18._____

19. $|9-8|$

19._____

20. $|25|-|7|$

20._____

21. $|5+4|+|7+8|$

21._____

22. $|50-42|+|32-29|$　　　　　　　　　　　　　**22.** _____

23. $|-14|-|9|$　　　　　　　　　　　　　　　　**23.** _____

24. $|26|-|-13|$　　　　　　　　　　　　　　　　**24.** _____

Name:
Instructor:

Date:
Section:

Chapter 1 ADDING AND SUBTRACTING INTEGERS AND POLYNOMIALS

1.6 Addition of Integers

Learning Objectives
A   Add integers with the same and with opposite signs.
B   Translate verbal expressions into mathematical expressions and simplify.
C   Identify properties of addition.

Getting Ready

Add or subtract the following whole numbers and decimals.

1. $8 + 17$ 1._____

2. $16 - 9$ 2._____

3. $9.1 - 4.6$ 3._____

4. Graph the following set of integers on the number line: $\{-4, -2, 0, 1, 3\}$.   4. See graph_____

Key Terms

Use the vocabulary terms listed below to complete each statement in exercises 1–7.

sum	associative	the same sign	opposite signs
additive identity		additive inverses	commutative

1. The _____ property of addition says the manner in which numbers are grouped does not matter.

2. The _____ property of addition says that the order in which numbers are added does not matter.

3. To add numbers with _____, add the absolute values of the numbers and give the answer the same sign as the numbers being added.

4. To add numbers with _____, subtract the absolute values of the numbers and give the answer the sign of the number with the larger absolute value.

Copyright © 2013 Pearson Education, Inc.

5. The word _____ indicates addition.

6. If the sum of two numbers is 0, the two numbers are _____ of each other.

7. The number 0 is called the _____.

Practice Problems

Objective A   Add integers with the same and with opposite signs.

Find the following sums.

1. $5 + 21$ 1._____

2. $-6 + 15$ 2._____

3. $-8 + (-23)$ 3._____

4. $7 + (-20)$ 4._____

5. $-11 + (-19)$ 5._____

6. $-31 + 17$ 6._____

7. $-4.2 + 5.1$ 7._____

8. $-5.6 + (-8.9)$ 8._____

9. $8 + 14 + (-7)$ 9._____

10. $-15 + 7 + (-26)$ 10._____

11. $[-7 + (-6)] + (-9 + 5)$ 11._____

Name:
Instructor:

Date:
Section:

12. $[15 + (-9)] + (-11 + 12)$

12._____

13. $-3 + (-15) + (-23)$

13._____

14. $15.2 + (-9) + 11.7$

14._____

Objective B   Translate verbal expressions into mathematical expressions and simplify.

Write a numerical expression for each of the following and evaluate.

15. The sum of −10 and 16

15._____

16. 18 increased by 7

16._____

17. 14 more than −9

17._____

18. −21 added to −19

18._____

Solve.

19. Kyra has a balance of $793 in her checking account. She writes a check for $137 and later makes a deposit of $381. What is her new balance?

19._____

20. Mandy is playing a card game where in the first round she earns 14 points, then earns another 18 points, then loses 7 points, then loses another 20 points. What is her score at the end of the round?

20._____

Copyright © 2013 Pearson Education, Inc.

Objective C Identify the properties of addition.

Give the name of the property illustrated by each of the following.

21. $-5+4=4+(-5)$ 21._____

22. $-8+8=0$ 22._____

23. $7+0=7$ 23._____

24. $(-4+6)+8=-4+(6+8)$ 24._____

25. $(-8+7)+6=6+(-8+7)$ 25._____

Complete each of the following using the given property.

26. $0+18=$ _____   additive identity   26._____

27. $-9+12=$ _____   commutative for addition 27._____

28. $5+(-19+18)=$ _____   associative for addition   28._____

29. $-30+30=$ _____   additive inverse   29._____

30. $11(6+3)=$ _____   commutative for addition   30._____

62   Copyright © 2013 Pearson Education, Inc.

Name: Date:
Instructor: Section:

Chapter 1 ADDING AND SUBTRACTING INTEGERS AND POLYNOMIALS

1.7 Subtraction of Integers and Combining Like Terms

Learning Objectives
A Subtract integers.
B Combine like terms.

Getting Ready

Add or subtract the following whole numbers and decimals:

1. $5.3 + 2.8$ 1._____

2. $26 - 18$ 2._____

Add the following integers:

3. $-6 + 9$ 3._____

4. $8 + (-13)$ 4._____

5. $-7 + (-8)$ 5._____

6. Find the negative (opposite) of -7.   6._____

Key Terms

Use the vocabulary terms listed below to complete each statement in exercises 1–5.

numerical coefficient   **variable**   **opposite**
term   **distributive property**

1. To subtract a number, add its _____.

2. A(n) _____ is a number, variable, or product and/or quotient of numbers and variables raised to powers.

3. The _____ states that for all numbers a, b, and c, $a(b+c) = a \cdot b + a \cdot c$.

4. In a term, the constant preceding the variable(s) is called the _____.

5. To add like terms, add the numerical coefficients and leave the _____ portion unchanged.

Practice Problems

Objective A   Subtract integers.

Evaluate each of the following.

1. $15 - 9$ 1._____

2. $9 - 15$ 2._____

3. $22 - 31$ 3._____

4. $-8 - 17$ 4._____

5. $16 - (-31)$ 5._____

6. $-9 - (-18)$ 6._____

7. $5 - (-7) - 11$ 7._____

8. $8 - (11 - 6)$ 8._____

9. $-19 - (20 - 8)$ 9._____

Name:
Instructor:

Date:
Section:

10. $24-(-7-14)$

10._____

11. $2.6-7.9$

11._____

12. $-2.6-9.9$

12._____

13. $14.2-7.8-2.9$

13._____

14. $(6-9)-(8-1)$

14._____

15. $(-45-8)-(6-11)$

15._____

16. $[(49-54)-17]-31$

16._____

17. $12-[5-(7-16)]$

17._____

18. $-52+[49-(-8+17)]$

18._____

Objective B Combine like terms.

Simplify the following by adding like terms.

19. $12x+6x$

19._____

20. $-7y+18y$

20._____

21. $4mn - 11mn + 8$ 21._____

22. $-2z + 14z - 6z$ 22._____

23. $4x^2 - 7y^2 - 15y^2 + 8x^2 + 10$ 23._____

24. $-9d^2 - 6 + 3d + 15d^2 - 17 - 14d$ 24._____

25. $a^2b + 7b - 4ab^2 + 18a^2b + 12b$ 25._____

Write each of the following as a subtraction problem and evaluate.

26. The difference of 4 and −7   26._____

27. 8 less 19   27._____

28. $5m$ less than $-m$ 28._____

29. The difference of $4ab$ and $9ab$ added to $18ab$ 29._____

30. −13 subtracted from 39   30._____

Name:
Instructor:
Date:
Section:

Chapter 1 ADDING AND SUBTRACTING INTEGERS AND POLYNOMIALS

1.8 Polynomial Definitions and Combining Polynomials

Learning Objectives
A   Identify types of polynomials.
B   Find the degree of a term.
C   Write polynomials in descending order and find their degrees.
D   Evaluate polynomials for a specified value of the variable.

Getting Ready

Substitute the given value for the variable and evaluate by using the order of operations.

1. $7x-5$: $x=4$

    1._____

2. y^2+5y-3: $y=2$

    2._____

Add the following like terms:

3. $6a+4-3a-5$

    3._____

4. $7p^2-2p-1+6p^2+8p+5$

    4._____

Translate the following expressions in English into mathematics and simplify.

5. $5x$ less than $-4x$

    5._____

6. $2a$ more than $-4a$

    6._____

Key Terms

Use the vocabulary terms listed below to complete each statement in exercises 1–7.

| polynomial | trinomial | monomial | binomial |
| algebraic expression | | term | degree |

1. A(n) _____ is any constant, variable, or combination of constants and variables using the operations of addition, subtraction, multiplication, division, raising to powers, or taking roots.

2. A(n) _____ is a number, variable, or product and/or quotient of numbers and variables raised to powers.

3. A polynomial with two terms is called a(n) _____ .

4. A polynomial with one term is called a(n) _____ .

5. A polynomial with three terms is called a(n) _____ .

6. A(n) _____ is the sum or difference of a finite number of terms in which there is no variable as a divisor and the exponents on the variables are whole numbers.

7. If a monomial has only one variable, the exponent on the variable is called the _____ of the monomial.

Practice Problems

Objective A Identify types of polynomials.

Determine whether each expression is a polynomial. If it is a polynomial, determine the number of terms. Identify any monomials, binomials, or trinominals.

1. $3x^2 + 7$ 1._____

2. $5 - 8x^3 + 7x$ 2._____

3. 14 3._____

4. $\dfrac{7y^3}{3x} - 17x$ 4._____

5. $9x^8 - 5x^3 + 15x^{12} - 17$ 5._____

Name:
Instructor:
Date:
Section:

Objective B Find the degree of a term.

Give the coefficient, variable(s), and degree of each of the following.

6. $9x$

6._____

7. $-8x^5$

7._____

8. $x^2 y^3 z^7$

8._____

9. $-y^{15} z$

9._____

Objective C Write polynomials in descending order and find their degrees.

Write each of the following in descending order and give the degree of the polynomial.

10. $3x - 5x^2 + 9$

10._____

11. $5x - 2x^3$

11._____

12. $1 - 4y + y^6 + 2y^5$

12._____

Objective D Evaluate polynomials for a specified value of the variable.

Evaluate each of the following for the indicated value of the variable.

13. $P(x) = 6x + 5;\ x = 2$

13._____

14. $P(y) = 3y^2 + 2y + 4;\ y = 7$

14._____

Copyright © 2013 Pearson Education, Inc.

Find the following sums and differences of polynomials.

15. $(4x-3y)+(5x+12y)$ 15._____

16. $(9x^2-8x+17)+(4x^2+6x-11)$ 16._____

17. $(8m-13n)-(14m+5n)$ 17._____

18. $(2a^2b+6a-5b+1)+(9a^2b-8b-17)$ 18._____

19. $(10p^2q^2+9p^2q-16pq^2-7p)-(-8p^2q^2+11p^2q+4p)$ 19._____

Represent each of the following using addition and/or subtraction and simplify.

20. Subtract $3x^2-8x+7$ from $9x^2+16x-11$. 20._____

21. Find the difference of $14y+3$ and $7y-6$. 21._____

22. From the sum of $8m^2-9m+2$ and $7m^2-6$ subtract $10m^2+8m+13$. 22._____

Name: Date:
Instructor: Section:

Chapter 2 LAWS OF EXPONENTS, PRODUCTS AND QUOTIENTS OF INETEGRS AND POLYNOMIALS

2.1 Multiplication of Integers

Learning Objectives
A Multiply signed numbers.
B Raise signed numbers to powers.
C Identify properties of multiplication.

Getting Ready

Find the products of the following whole numbers:

1. $(8)(6)(2)$ 1._____

2. $(7)(5)(3)(4)$ 2._____

Raise the following whole numbers to powers:

3. 3^5 3._____

4. 2^6 4._____

Evaluate the following expressions for the given value(s) of the variable(s):

5. $4ab$: $a=5$, $b=4$ 5._____

6. $x^3 y^2$: $x=4$, $y=2$ 6._____

Key Terms

Use the vocabulary terms listed below to complete each statement in exercises 1–7.

      inverse   negative   positive   commutative property of multiplication
      identity   product   associative property of multiplication

1. When multiplying numbers, we are finding their _____.

Copyright © 2013 Pearson Education, Inc.   71

2. The _____ states that multiplication may be performed in any order.

3. The product of a negative number and a positive number is _____.

4. The _____ states that multiplication may be grouped in any manner.

5. The number 1 is the _____ for multiplication.

6. The product of a negative number and a negative number is _____.

7. The multiplicative _____ of the number a, $a \neq 0$, is $\frac{1}{a}$.

Practice Problems

Objective A Multiply signed numbers.

Find the following products.

1. $6(-3)$ 1._____

2. $(-7)(-11)$ 2._____

3. $2(5)(-8)$ 3._____

4. $(-1.2)(-2.3)$ 4._____

5. $(-4)(-6)(-5)$ 5._____

Objective B Raise signed numbers to powers.

Evaluate.

6. -7^2 6._____

7. $(-3)^2$ 7._____

Name:
Instructor:
Date:
Section:

8. $(-5)^4$
8._____

9. $(-6)^3$
9._____

10. $(-4)^2(-3)^3$
10._____

11. $(6)^2(-5)^2$
11._____

Evaluate the following for $x = 2$ and $y = -5$.

12. $4xy$
12._____

13. $-3x^3y$
13._____

14. $-x^4y^2$
14._____

Write an expression for each of the following and evaluate.

15. The product of 2 and -9 increased by 7
15._____

16. -9 decreased by the product of -6 and 8
16._____

17. One-half of the difference of 8 and -12
17._____

Objective C Identify properties of multiplication.

Identify the property illustrated by each of the following.

18. $7(-8+9) = (-8+9)7$ 18. _____

19. $7 \cdot \dfrac{1}{7} = 1$ 19. _____

20. $10 \cdot 0 = 0$ 20. _____

Complete each of the following by using the given property of multiplication.

21. $-4[(5)(-12)] = $ _____ associative 21. _____

22. $(-8)(7) = $ _____ commutative 22. _____

23. $(-10) \cdot 1 = $ _____ identity for multiplication 23. _____

Name:
Instructor:

Date:
Section:

Chapter 2 LAWS OF EXPONENTS, PRODUCTS AND QUOTIENTS OF INTEGERS AND POLYNOMIALS

2.2 Multiplication Laws of Exponents

Learning Objectives
A   Simplify expressions of the form $a^m \cdot a^n$.
B   Simplify expressions of the form $(ab)^n$.
C   Simplify expressions of the form $(a^m)^n$.
D   Simplify expressions involving two or more of the preceding forms A through C.

Getting Ready

Add or multiply the following whole numbers:

1. $8 + 6$ 1._____

2. $2 + 4 + 9$ 2._____

3. $2 \cdot 8$ 3._____

4. $4 \cdot 9$ 4._____

Add the following like terms:

5. $5a^2 + 7a^2$ 5._____

6. $19x^3y - 11x^3y$ 6._____

Key Terms

Use the vocabulary terms listed below to complete each statement in exercises 1–3.

       add   **factor**   **multiply**

1. To raise a product to a power, raise each _____ to the power.

2. To multiply two expressions with the same base, leave the base unchanged and _____the exponents.

3. To raise a number to a power to another power, leave the base unchanged and _____ the powers.

Practice Problems

Objective A Simplify expressions of the form $a^m \cdot a^n$.

Find the following products.

1. $t^8 \cdot t^3$ 1._____

2. $6^3 \cdot 6^5$ 2._____

3. $a^{10} \cdot a^5 \cdot a^7$ 3._____

Simplify.

4. $(8x)(3x)+(5x)(-9x)$ 4._____

5. $(-4y)(-7y^2)-(3y^2)(-2y)$ 5._____

Objective B Simplify expressions of the form $(ab)^n$.

Simplify the following using the laws of exponents and leave answers in exponential form.

6. $(xy)^6$ 6._____

7. $(-3d)^4$ 7._____

8. $-(-2a)^5$ 8._____

9. $(2abc)^4$ 9._____

10. $(-4mn)^3$ 10._____

Name:
Instructor:

Date:
Section:

Objective C Simplify expressions of the form $(a^m)^n$.

Simplify the following using the laws of exponents and leave answers in exponential form.

11. $(m^{10})^6$ 　　　　　　　　　　　　　　　　　　　11._____

12. $(3^8)^{10}$ 　　　　　　　　　　　　　　　　　　12._____

13. $(x^8)^4$ 　　　　　　　　　　　　　　　　　　　13._____

14. $(5^3)^3$ 　　　　　　　　　　　　　　　　　　　14._____

Objective D Simplify expressions involving two or more of the preceding forms A through C.

Simplify the following using the laws of exponents and leave answers in exponential form.

15. $(5d^2)^3$ 　　　　　　　　　　　　　　　　　　15._____

16. $(xy)^4(xy)^8$ 　　　　　　　　　　　　　　　　16._____

17. $(-2^4 x^3)^2$ 　　　　　　　　　　　　　　　　17._____

18. $(a^2 b^3)(a^3 b)^4$ 　　　　　　　　　　　　　　18._____

19. $(-2p^2 q)^4 (3p^3 q^5)^2$ 　　　　　　　　　　　19._____

Copyright © 2013 Pearson Education, Inc.

20. $\left(-4^3 y^8\right)^5$

20._____

21. $\left(7^3 m^6\right)^4 \left(7^2 m^4\right)^3$

21._____

22. $\left(8a^2 b^5 c\right)^2$

22._____

23. $-\left(9^2 b^5\right)^2$

23._____

24. $\left(m^2 n^3\right)^4 \left(m^6 n^2\right)^5$

24._____

25. $-\left(5^2 x^7\right)^6$

25._____

26. $\left(a^2 b^5 c^9\right)^4 \left(ab^3 c^6\right)^2$

26._____

Simplify the following

27. $(4x)^2 + 6x(-3x)$

27._____

28. $(-5y)^3 - (6y)(7y^2)$

28._____

29. $(2n^3)^2 + (-8n^2)^3$

29._____

30. $(3xy^3)^3 - (2xy)^2(5xy^7)$

30._____

Name: Date:
Instructor: Section:

Chapter 2   LAWS OF EXPONNETS, PRODUCTS AND QUOTIENTS OF INTEGERS AND POLYNOMIALS

2.3   Products of Polynomials

Learning Objectives
A   Find the product of a monomial and a polynomial.
B   Find the product of two polynomials.

Getting Ready

1. Simplify $-6(4-8)$ by using the distributive property.   1._____

2. Simplify $4x^3 y(-3xy^5)$ by using $a^m \cdot a^n = a^{m+n}$.   2._____

3. Multiply the integers: $8(-5)$.   3._____

4. Simplify $2x^2 + 6x + 3x^2 - 4x + 7$ by adding like terms.   4._____

5. Find the area of the rectangle whose width is 7 inches and whose length is 11 inches.   5._____

Key Terms

Use the vocabulary terms listed below to complete each statement in exercises 1–3.

     rectangle   **triangle**   **distributive property**

1. The formula for area of a _____ is $A = \dfrac{bh}{2}$.

2. The _____ is used when multiplying a monomial and a polynomial.

3. The formula for area of a _____ is $A = LW$.

Practice Problems

Objective A Find the product of a monomial and a polynomial.

Find the products of the following monomials and polynomials..

1. $7(2a-3b)$ 1._____

2. $-6(5x+3y)$ 2._____

3. $-4(-2m+5n)$ 3._____

4. $3x^3(5x^4-7)$ 4._____

5. $-6y^4(2y^3-8)$ 5._____

6. $8(7b^2-5b+1)$ 6._____

7. $-t(-2t^6+4t^4+9)$ 7._____

8. $-4h^3(3h^6-5h^2-7)$ 8._____

9. $2a^3b^4(9a^2b^6+5a^4b^8)$ 9._____

10. $-5c^2d^7(-4cd^3+2c^2d^5-6c^8d)$ 10._____

Simplify the following.

11. $7(x+1)+8(x-2)$ 11._____

12. $-5(a-9)-4(a+3)$ 12._____

Copyright © 2013 Pearson Education, Inc.

Name:
Instructor:

Date:
Section:

13. $3y(2y+1)+5(y-7)$ 13._____

14. $-z(4z+1)+7(2z-8)$ 14._____

15. $5m(2m-7n)+6n(3m-2n)$ 15._____

16. $-8b(-5b+6c)-2b(b-3c)$ 16._____

17. $2(6a^2-5a+7)+3(4a^2+a-8)$ 17._____

18. $7(3x^2-2x-4)-5(2x^2+3x-6)$ 18._____

Objective B Find the product of two polynomials.

Find the products of the following polynomials.

19. $(y+3)(y+7)$ 19._____

20. $(x-9)(x+4)$ 20._____

21. $(w-8)(w-9)$ 21._____

22. $(2a+3)(a-6)$ 22._____

23. $(5b-4)(2b+1)$ 23._____

24. $(5m-6)(4m-3)$ 24._____

25. $(b-5)(b+6)$ 25._____

26. $(2n+3)(2n+5)$ 26._____

27. $(x+1)(2x^2-3x-7)$ 27._____

28. $(5a-2)(4a^2+a-7)$ 28._____

29. $8x+3$ 29._____
$2x-1$

30. $7x-5$ 30._____
$4x-3$

Write an expression for the area of each of the following using the given dimensions. Recall that the formula for the area of a rectangle is $A = LW$ and the formula for the area of a triangle is $A = \dfrac{bh}{2}$.

31. A rectangle with $L = x+2$ and $W = 3x+5$ 31._____

32. A rectangle with $L = 2x-7$ and $W = 5x+1$ 32._____

33. A triangle with $b = 3x$ and $h = 8x$ 33._____

34. A triangle with $b = 8y$ and $h = 4y$ 34._____

Name:
Instructor:
Date:
Section:

Chapter 2 LAWS OF EXPONENTS, PRODUCTS AND QUOTIENTS OF INTEGERS AND POLYNOMIALS

2.4 Special Products

Learning Objectives
A Multiply binomials using FOIL.
B Find products of the form $(a+b)(a-b)$.
C Square binomials.
D Optional: Recognize products of the form $(a+b)(a^2-ab+b^2)$ and $(a-b)(a^2+ab+b^2)$.

Getting Ready

1. Simplify $y(4y+5)y$ by using the distributive property.   1._____

2. Add the like terms: $-4x+7x$.   2._____

3. Simplify $-7c \cdot 6c$, using $a^m \cdot a^n = a^{m+n}$.   3._____

4. Multiply the integers: $-5(-3)$.   4._____

5. Raise the following integer to the indicated power: $(-6)^2$.   5._____

Key Terms

Use the vocabulary terms listed below to complete each statement in exercises 1–3.

    difference of squares   **FOIL**   **conjugate pairs**

1. The sum and difference of the same terms are referred to as _____.

2. The product of two binomials that are the sum and difference of the same terms results in a binomial that is a _____.

3. The word _____ gives us an easy way to remember which terms to multiply when finding the product of two binomials.

Copyright © 2013 Pearson Education, Inc.

Practice Problems

Objective A Multiply binomials using FOIL.

Find the product of the following binomials using FOIL.

1. $(x+7)(x+4)$

    1._____

2. $(y-8)(y+2)$

    2._____

3. $(w+3)(w-4)$

    3._____

4. $(a-7)(a-6)$

    4._____

5. $(6b+5)(3b-4)$

    5._____

6. $(9m+n)(2m-3n)$

    6._____

7. $(2x+3y)(3x+2y)$

    7._____

8. $(4c-5d)(4c+3d)$

    8._____

Objective B Find products of the form $(a+b)(a-b)$.

Find the following products of the form $(a+b)(a-b)$.

9. $(x+8)(x-8)$

    9._____

10. $(y-10)(y+10)$

    10._____

11. $(4z+5)(4z-5)$

    11._____

Name:
Instructor:

Date:
Section:

12. $(3a-b)(3a+b)$ 12._____

13. $(6c+7d)(6c-7d)$ 13._____

14. $(9-8a)(9+8a)$ 14._____

Objective C Square binomials.

Square the following binomials.

15. $(x+11)^2$ 15._____

16. $(y-1)^2$ 16._____

17. $(2w-5)^2$ 17._____

18. $(1+4n)^2$ 18._____

19. $(4x+5y)^2$ 19._____

20. $(2m^2+n)^2$ 20._____

21. $(6a^2+3b)^2$ 21._____

22. $(7w+8)^2$ 22._____

Find the following products.

23. $(x+14)(x-14)$ 23._____

Copyright © 2013 Pearson Education, Inc.

24. $(2y+7)(3y+5)$ 24._____

25. $(6a-b)^2$ 25._____

26. $(3z+8w)(3z-8w)$ 26._____

27. $(2m-7n)(7m-2n)$ 27._____

28. $(c-9d)^2$ 28._____

Objective D Optional: Recognize products of the form $(a+b)(a^2-ab+b^2)$ and $(a-b)(a^2+ab+b^2)$.

Find the following products.

29. $(a+1)(a^2-a+1)$ 29._____

30. $(b-3)(b^2+3b+9)$ 30._____

31. $(a-7)(a^2+7a+49)$ 31._____

32. $(x+y)(x^2-xy+y^2)$ 32._____

33. $(2a+5b)(4a^2-10ab+25b^2)$ 33._____

34. $(4z-3w)(16z^2+12zw+9w^2)$ 34._____

Name: Date:
Instructor: Section:

Chapter 2 LAWS OF EXPONENTS, PRODUCTS AND QUOTIENTS OF INTEGERS AND POLYNOMIALS

2.5 Division of Integers and Order of Operations with Integers

Learning Objectives
A Divide integers.
B Evaluate expressions containing integers that involve combinations of addition, subtraction, multiplication, division and raising to powers.
C Evaluate expressions containing variables when given the value(s) of the variables.

Getting Ready

1. Divide the whole numbers: $\frac{72}{8}$. 1._____

Add, subtract, or multiply the following integers:

2. $4-(-9)$ 2._____

3. $-6(8)$ 3._____

4. Raise the following integer to the indicated power: 2^7. 4._____

5. Simplify $-6+2 \cdot 12 \div 3 \cdot 6$ by using the order of operations. 5._____

Substitute the following value(s) for the variable(s) and evaluate:

6. $2x^2 - 4y$: $x=5$, $y=3$ 6._____

7. $\frac{3a^2 + 5b}{a^3 - b^2}$: $a=3$, $b=5$ 7._____

Key Terms

Use the vocabulary terms listed below to complete each statement in exercises 1–5.

multiplication   positive   negative   zero   quotient

1. The quotient of two numbers with opposite signs is _____.

2. Division by _____ is undefined.

3. The word _____ indicates the operation of division.

4. The quotient of two numbers with the same sign is _____.

5. Division can always be checked by _____.

Practice Problems

Objective A Divide integers.

Find the following quotients.

1. $\dfrac{27}{-9}$

   1._____

2. $\dfrac{-27}{-9}$

   2._____

3. $\dfrac{-27}{9}$

   3._____

4. $\dfrac{15}{0}$

   4._____

5. $\dfrac{0}{15}$

   5._____

6. $\dfrac{-42}{21}$

   6._____

7. $\dfrac{126}{-6}$

   7._____

8. $\dfrac{-67}{-1}$

   8._____

Name: Date:
Instructor: Section:

Write an expression for each of the following and simplify.

9. The quotient of 88 and −8 increased by 9

9._____

10. 5 less the quotient of 24 and −8

10._____

11. The quotient of −18 and −9, subtracted from −10

11._____

12. The quotient of −30 and 15 subtracted from the quotient of 20 and −4

12._____

Objective B Evaluate expressions containing integers that involve combinations of addition, subtraction, multiplication, division and raising to powers.

Evaluate the following.

13. $-4-6+7$

13._____

14. $9-7+18$

14._____

15. $12-4 \cdot 8$

15._____

16. $2-9 \cdot 4$

16._____

17. $8-66 \div (-11)$

17._____

18. $3 - 21 \div (-7)$ 18._____

19. $16 - 2(-8)^2$ 19._____

20. $14 + 2(-7)^2$ 20._____

21. $8 + 3 \cdot 4^2 \div (-6)(-7) + 9$ 21._____

22. $-7 - 2 \cdot 6^2 \div 9(-8) - 15$ 22._____

23. $-7^2 (10 - 3^2)$ 23._____

24. $|5 \cdot 4^2 - 19| - |8(3-7)|$ 24._____

25. $|12 - 4(7-5) \div 2| + |6(1 - 3^2) \div 4|$ 25._____

26. $\dfrac{9^2 - 15}{6^2 - 3}$ 26._____

Name:
Instructor:

Date:
Section:

27. $\dfrac{-4(-6)^2 - 12}{3(-2)^2 + 3(3)^2}$

27._____

28. $\dfrac{\left|7(6-4^2)\right|}{\left|3\cdot 2^2 - 6^2 \cdot 4\right|}$

28._____

Objective C Evaluate expressions containing variables when given the value(s) of the variables.

Evaluate each of the following for $x = -4$, $y = 5$, *and* $z = -2$.

29. $3y^2 - 5x$

29._____

30. $-2z^3 + 4y$

30._____

31. $6x^2 yz$

31._____

32. $-8xz^3$

32._____

33. $-4z + x(5y - 7x)$

33._____

Copyright © 2013 Pearson Education, Inc.

34. $\dfrac{x^2 - y^2}{z^2 - x^2}$

34._____

Write an expression for each of the following using signed numbers and evaluate.

35. The Shirt Shack sold 150 blazers for a profit of $6.50 each and 88 shirts at a loss of $2.00 each. What was their net profit/loss on the sale of blazers and shirts?

35._____

36. Andrea played a game that consisted of 10 rounds. She scored 15 points in each of 3 rounds, lost 10 points in each of 2 rounds and lost 5 points in each of 5 rounds. What was her final score?

36._____

Name:
Instructor:

Date:
Section:

Chapter 2 LAWS OF EXPONENTS, PRODUCTS AND QUOTIENTS OF INTEGERS AND POLYNOMIALS

2.6 Quotient Rule and Integer Exponents

Learning Objectives

A Simplify exponential expressions using the property $\dfrac{a^m}{a^n} = a^{m-n}$.

B Simplify expressions with zero exponents.

C Simplify expressions with negative integer exponents.

Getting Ready

Add or subtract the following integers:

1. $6 - 9$ 1._____

2. $-4 - (-7)$ 2._____

Simplify the following by using $\left(a^m\right)^n = a^{mn}$ or $(ab)^n = a^n b^n$:

3. $\left(2^4\right)^3$ 3._____

4. $(4x)^3$ 4._____

Raise the following integers to powers:

5. 8^2 5._____

6. $(-5)^3$ 6._____

Key Terms

Use the vocabulary terms listed below to complete each statement in exercises 1–5.

    base   one   exponent   zero   reciprocals

1. Any nonzero number raised to the _____ power is equal to 1.

2. Any number (other than 0) divided by itself is equal to _____ .

3. The _____ indicates how many times the base is to be used as a factor.

4. To divide two numbers with the same base, take the top exponent and subtract the bottom exponent, leaving the _____ unchanged.

5. x^n and x^{-n} are _____ .

Practice Problems

Objectives A, B, C Simplify exponential expressions using the property $\dfrac{a^m}{a^n} = a^{m-n}$. Simplify expressions with zero exponents. Simplify expressions with negative integer exponents.

Simplify each expression. Express answers in exponential form with positive exponents only. Assume that all variables represent nonzero quantities.

1. $\dfrac{3^5}{3^2}$ 　　　　　　　　　　　　　　　　　　　　　1._____

2. $\dfrac{2^9}{2^5}$ 　　　　　　　　　　　　　　　　　　　　　2._____

3. $\dfrac{x^{12}}{x^9}$ 　　　　　　　　　　　　　　　　　　　　3._____

4. $\dfrac{y^{19}}{y^8}$ 　　　　　　　　　　　　　　　　　　　　4._____

5. $\dfrac{z^{11}}{z^{10}}$ 　　　　　　　　　　　　　　　　　　　5._____

6. $\dfrac{7^4}{7}$ 　　　　　　　　　　　　　　　　　　　　　6._____

7. $\dfrac{4^6}{4^6}$ 　　　　　　　　　　　　　　　　　　　　　7._____

Name: Date:
Instructor: Section:

8. $\dfrac{d^5}{d^5}$ 8._____

9. 12^0 9._____

10. m^0 10._____

11. $13y^0$ 11._____

12. $(14a)^0$ 12._____

13. $(7m)^0 + 8m^0$ 13._____

14. $(-21z)^0$ 14._____

15. y^{-7} 15._____

16. $\dfrac{1}{x^{-8}}$ 16._____

17. $(4a)^{-3}$ 17._____

18. $5b^{-4}$ 18._____

19. $-8d^{-3}$ 19._____

20. $x^{-9}x^8$

20._____

21. $a^{-7}a^{-8}$

21._____

22. $\left(m^{-7}\right)^8$

22._____

23. $\left(z^{-2}\right)^{-6}$

23._____

24. $\dfrac{7^{-2}}{7^9}$

24._____

25. $\dfrac{z^6}{z^{-18}}$

25._____

26. $\left(15^{-3}\right)^{-5}$

26._____

27. $\left(4^{-1}\right)^9$

27._____

28. $b^{-4} \cdot b^0$

28._____

29. $\dfrac{q^4}{q^6}$

29._____

30. $\dfrac{2}{y^{-9}}$

30._____

Name:
Instructor:
Date:
Section:

Chapter 2 LAWS OF EXPONENTS, PRODUCTS AND QUOTIENTS OF INTEGERS AND POLYNOMIALS

2.7 Power Rule for Quotients and Using Combined Laws of Exponents

Learning Objectives

A Simplify exponential expressions by using the property $\left(\dfrac{a}{b}\right)^n = \dfrac{a^n}{b^n}$.

B Simplify exponential expressions by using the property $\left(\dfrac{a}{b}\right)^{-n} = \left(\dfrac{b}{a}\right)^n$.

C Simplify expressions that involve integer exponents by using more than one law of exponents.

Getting Ready

1. Find the following power of the integer: $(-3)^5$.

1._____

Add, subtract, or multiply the following integers:

2. $-5+11$

2._____

3. $4(-6)$

3._____

Simplify the following, using $a^m \cdot a^n = a^{m+n}$, $(ab)^n = a^n b^n$, *and* $\left(a^m\right)^n = a^{mn}$:

4. $\left(3x^2 y\right)^3$

4._____

5. $\left(a^2\right)^4 \left(b^3\right)^5$

5._____

Simplify the following, using $\dfrac{a^m}{a^n} = a^{m-n}$, $x^0 = 1$, *and* $x^{-n} = \dfrac{1}{x^n}$:

6. $\dfrac{y^6}{y^6}$

6._____

7. $\dfrac{c^3}{c^7}$

7._____

Copyright © 2013 Pearson Education, Inc.

Key Terms

Use the vocabulary terms listed below to complete each statement in exercises 1–11.

a^{m+n}	$a^n b^n$	a^{mn}	a^{m-n}	a^n	factors
$\dfrac{a^n}{b^n}$	$\left(\dfrac{b}{a}\right)^n$	$\dfrac{1}{a^n}$	quotient	invert	

1. To raise a product to a power, you raise each of the _____ to the power.

2. $\left(\dfrac{a}{b}\right)^n =$ _____.

3. To raise a _____ to a power, raise both the numerator and denominator to the power.

4. $a^m a^n =$ _____.

5. $\dfrac{1}{a^{-n}} =$ _____.

6. $(ab)^n =$ _____.

7. To raise a fraction to a negative power, _____ the fraction and change the power to positive.

8. $a^{-n} =$ _____.

9. $\left(a^m\right)^n =$ _____.

10. $\left(\dfrac{a}{b}\right)^{-n} =$ _____.

11. $\dfrac{a^m}{a^n} =$ _____.

Name:
Instructor:

Date:
Section:

Practice Problems

Objective A Simplify exponential expressions by using the property $\left(\dfrac{a}{b}\right)^n = \dfrac{a^n}{b^n}$.

Write each of the following with positive exponents only. Assume that all variables represent nonzero quantities.

1. $\left(\dfrac{x}{y}\right)^2$

   1._____

2. $\left(\dfrac{3}{z}\right)^3$

   2._____

3. $\left(\dfrac{5}{d}\right)^4$

   3._____

4. $\left(\dfrac{k}{7}\right)^3$

   4._____

5. $\left(\dfrac{6}{5}\right)^2$

   5._____

6. $\left(\dfrac{m}{n}\right)^{12}$

   6._____

Objective B Simplify exponential expressions by using the property $\left(\dfrac{a}{b}\right)^{-n} = \left(\dfrac{b}{a}\right)^n$.

Write each of the following with positive exponents only. Assume that all variables represent nonzero quantities.

7. $\left(\dfrac{m}{n}\right)^{-4}$

   7._____

Copyright © 2013 Pearson Education, Inc.

8. $\left(\dfrac{w}{z}\right)^{-6}$ 8._____

9. $\left(\dfrac{2}{p}\right)^{-3}$ 9._____

10. $\left(\dfrac{a}{4}\right)^{-3}$ 10._____

11. $\left(\dfrac{x}{y}\right)^{-12}$ 11._____

12. $\left(\dfrac{7}{6}\right)^{-2}$ 12._____

Objective C   Simplify expressions that involve integer exponents by using more than one law of exponents.

Write each of the following with positive exponents only. Assume that all variables represent nonzero quantities.

13. $\left(\dfrac{3r}{s}\right)^{4}$ 13._____

14. $\left(\dfrac{p^2}{2q^5}\right)^{6}$ 14._____

15. $\left(\dfrac{c^4}{d^{11}}\right)^{8}$ 15._____

Name:
Instructor:

Date:
Section:

16. $\left(\dfrac{x^{-3}}{y^2}\right)^{-5}$

16._____

17. $\left(\dfrac{w^{-5}}{z^{-7}}\right)^{-4}$

17._____

18. $\left(\dfrac{9^0 a^5}{3a^3}\right)^4$

18._____

19. $\left(-6x^{-5}\right)^{-3}$

19._____

20. $\left(8m^{-9}n^6\right)^{-2}$

20._____

21. $\dfrac{\left(x^4\right)^5 \left(x^5\right)^4}{\left(x^2\right)^9}$

21._____

22. $\dfrac{\left(5d^{-3}\right)^4}{d^6 d^{10}}$

22._____

23. $\dfrac{\left(12p^6\right)^0}{\left(5p^3\right)^2}$

23._____

Copyright © 2013 Pearson Education, Inc.

24. $\dfrac{(2x^7)^{-3}(3x^5)^{-2}}{(4x^6)^{-3}}$ 24._____

25. $\dfrac{(5m^4)(7m^6)}{(m^8)(m^8)}$ 25._____

26. $(4c^5)^{-4}$ 26._____

27. $(-9z^{-5})^{-3}$ 27._____

28. $\left(\dfrac{c^{14}}{d^{18}}\right)^4$ 28._____

29. $\left(\dfrac{2w^{-3}}{v^5}\right)^4$ 29._____

30. $\dfrac{(7b^3)^2}{(7^0 b^4)^5}$ 30._____

Name:
Instructor:
Date:
Section:

Chapter 2 LAWS OF EXPONENTS, PRODUCTS AND QUOTIENTS OF INTEGERS AND POLYNOMIALS

2.8 Division of Polynomials by Monomials

Learning Objectives
A   Divide monomials by monomials.
B   Divide polynomials of more than one term by monomials.

Getting Ready

1. Divide the following integers: $\dfrac{-42}{-7}$.

    1._____

Simplify the following, using $\dfrac{a^m}{a^n} = a^{m-n}$ and $x^{-n} = \dfrac{1}{x^n}$:

2. $4x^{-4}$

    2._____

3. $\dfrac{20a^5}{-4a^3}$

    3._____

4. Subtract the following integers: $-7-(-9)$.

    4._____

5. Simplify $\left(6x^3 y^5\right)^3$, using $(ab)^n = a^n b^n$ and $\left(a^m\right)^n = a^{mn}$.

    5._____

Key Terms

Use the vocabulary terms listed below to complete each statement in exercises 1–2.

    monomial   **term**

1. To divide a polynomial by a monomial, divide each _____ of the polynomial by the monomial.

2. A polynomial with one term is called a _____.

Practice Problems

Objective A Divide monomials by monomials.

Find the quotient of the following monomials.

1. $\dfrac{7y^4}{y}$

   1._____

2. $\dfrac{-6x^5}{x^2}$

   2._____

3. $\dfrac{8d^9}{2d^6}$

   3._____

4. $\dfrac{32t^8}{-8t^3}$

   4._____

5. $\dfrac{c^{10}d^6}{c^4d^2}$

   5._____

6. $\dfrac{m^8n^7}{-m^5n^3}$

   6._____

7. $\dfrac{-58xy^5}{2x^6y^2}$

   7._____

8. $\dfrac{-38x^5y^6z^9}{-19x^2y^{12}z}$

   8._____

9. $\dfrac{\left(4a^3b^6\right)^4}{8a^6b^8}$

   9._____

Name:
Instructor:

Date:
Section:

10. $\dfrac{-\left(7m^6n^3\right)^2}{7m^{15}n^4}$

10._____

11. $\dfrac{14vw^9}{\left(7vw^3\right)^2}$

11._____

12. $\dfrac{18a^7b^{11}c^{20}}{9a^{10}b^{12}c}$

12._____

Objective B   Divide polynomials of more than one term by monomials.

Find the quotient of the following polynomials and monomials.

13. $\dfrac{8x+24}{4}$

13._____

14. $\dfrac{y^3+y^2}{y}$

14._____

15. $\dfrac{8m^6-4m}{4m}$

15._____

16. $\dfrac{a^5-a^4-a}{a}$

16._____

17. $\dfrac{12c^5d^6-24c^7d^8}{6c^2d^3}$

17._____

18. $\dfrac{14y^4-28y^3+56y}{7y^2}$

18._____

19. $\dfrac{6z^5 - 8z}{3z^3}$ 19._____

20. $\dfrac{36pq^5 + 42p^6q^3}{6p^4q^2}$ 20._____

21. $\dfrac{12x^3y^4 - 60x^2y^7 + 6x^9y^8}{3x^3y^2}$ 21._____

22. $\dfrac{15a^2 - 30a + 5}{5}$ 22._____

23. $\dfrac{m^4 + m^6 + m^8}{m}$ 23._____

24. $\dfrac{16w^4z^8 + 24w^3z^9 + 12w^6z}{4w^3z^2}$ 24._____

25. $\dfrac{80uv^7 + 70u^6v^9 - 40u^2v}{10uv}$ 25._____

Write an expression for each of the following and simplify.

26. Find the quotient of $20a^5b^6$ and $4a^2b^3$. 26._____

27. Divide $8x^6 - 16x^5 + 12x^3$ by $4x^2$. 27._____

28. Divide the sum of $9m^3n^4$ and $-15m^6n^8$ by $3mn^2$. 28._____

29. Find the quotient of $-33y^8z^{11}$ and $-11y^4z^{13}$. 29._____

30. Divide the sum of $7x^6 + 5x^4 - x^2$ and $8x^6 - 4x^4 + 3x^2$ by $3x^3$. 30._____

Name: Date:
Instructor: Section:

Chapter 2 LAWS OF EXPONNETS, PRODUCTS AND QUOTIENTS OF INTEGERS AND POLYNOMIALS

2.9 An Application of Exponents: Scientific Notation

Learning Objectives
A   Change numbers written in scientific notation to numbers in standard notation.
B   Write numbers in scientific notation.
C   Multiply and divide very large and/or very small numbers using scientific notation.
D   Solve problems by using scientific notation.

Getting Ready

1. Add or subtract the following integers: $-3+11$.   1._____

Multiply or divide the following whole numbers and decimals:

2. $(4.1)(3.5)$ 2._____

3. $\dfrac{18.165}{5.25}$ 3._____

4. Simplify $10^8 \cdot 10^{-3}$ by using $a^m \cdot a^n = a^{m+n}$. Leave the answer as a power of 10.   4._____

Simplify, by using $\dfrac{a^m}{a^n} = a^{m-n}$. Leave the answers as a power of 10.

5. $\dfrac{10^{-8}}{10^6}$ 5._____

6. $\dfrac{10^{-4}}{10^{-9}}$ 6._____

Copyright © 2013 Pearson Education, Inc.

Key Terms

Use the vocabulary terms listed below to complete each statement in exercises 1–5.

positive   negative   exponential part
coefficient   scientific notation

1. A number is written in _____ if it is in the form of $a \times 10^n$, where $1 \leq a < 10$ and n is an integer.

2. To find the product of the form $a \times 10^n$, move the decimal n places to the right if n is _____.

3. The "a" of $a \times 10^n$ is referred to as the_____.

4. To find the product of the form $a \times 10^n$, move the decimal $|n|$ places to the left if n is _____.

5. The "10^n" of $a \times 10^n$ is referred to as the_____.

Practice Problems

Objective A Change numbers written in scientific notation to numbers in standard notation.

Write the following in standard notation.

1. 2.6×10^5 1._____

2. 4.9×10^3 2._____

3. 3.2×10^{-4} 3._____

4. 8×10^{-3} 4._____

5. 9×10^7 5._____

6. 1×10^{-7} 6._____

Name: Date:
Instructor: Section:

Objective B   Write numbers in scientific notation.

Write each of the following in scientific notation.

7. 62,000   7._____

8. 1,540,000,000   8._____

9. .0000026   9._____

10. .00051   10._____

11. 937,000   11._____

12. .004   12._____

13. The net worth of a company is $26,750,000,000.   13._____
Write the net worth in scientific notation.

14. The density of silver is approximately 10,500   14._____
kilograms per cubic meter. Express the density of
silver in scientific notation.

Objective C   Multiply and divide very large and/or very small numbers using scientific notation.

Simplify by using scientific notation. Express answers in both scientific and standard notation.

15. $(2.3 \times 10^4)(1.2 \times 10^6)$ 15._____

16. $(4.1 \times 10^8)(5.2 \times 10^{-6})$ 16._____

17. $(1.05 \times 10^{-2})(7.5 \times 10^{-9})$ 17._____

18. $(6.43 \times 10^4)(2.7 \times 10^9)$ 18._____

19. $\dfrac{2.6 \times 10^9}{1.3 \times 10^5}$ 19._____

20. $\dfrac{4.8 \times 10^{-6}}{1.6 \times 10^6}$ 20._____

21. $\dfrac{(2.25 \times 10^{-4})(4 \times 10^9)}{1.5 \times 10^{-3}}$ 21._____

22. $\dfrac{(1.2 \times 10^8)(2.0 \times 10^6)}{(3 \times 10^4)(1.6 \times 10^7)}$ 22._____

Name: Date:
Instructor: Section:

Simplify by using scientific notation. Express answers in both scientific and standard notation.

23. $(75,000)(4600)$ **23.** _____

24. $(1,240,000)(2,600,000)$ **24.** _____

25. $(.000015)(35,000)$ **25.** _____

26. $(.0000009)(.0000085)$ **26.** _____

27. $\dfrac{21,600}{3,000,000}$ **27.** _____

28. $\dfrac{.000016}{.000000008}$ **28.** _____

29. $\dfrac{(240,000)(.0012)}{(.000006)(3200)}$ **29.** _____

30. $\dfrac{(9,000,000)(84,000)}{(3000)(420,000)}$ **30.** _____

Objective D Solve problems by using scientific notation.

Solve.

31. The approximate distance from Earth to the planet Venus is 25 million miles. How many hours would it take a spacecraft traveling at 22,500 miles per hour to reach Venus?

31._____

32. If a computer can execute a command in .00004 second, how long will it take the computer to execute 7 million commands?

32._____

33. The solubility product of silver acetate is 1.82×10^{-3} and the solubility product of nickel sulfide is 1.4×10^{-24}. How many times greater is the solubility product of silver acetate than that of nickel sulfide? State your answer in scientific notation.

33._____

34. The speed of light is approximately 3×10^{10} cm per second. How long will it take light to travel 15×10^{18} cm?

34._____

Name: Date:
Instructor: Section:

Chapter 3 LINEAR EQUATIONS AND INEQUALITIES

3.1 Addition Property of Equality

Learning Objectives
A Solve linear equations, using the addition property of equality.
B Translate mathematics expressions into English and English expressions into mathematics.
C Solve various types of real-world problems.

Getting Ready

Add or subtract as indicated.

1. $-6+4$ 1._____

2. $-9-3$ 2._____

3. $-4.1+5.3$ 3._____

4. $-6.7-4.2$ 4._____

Perform the indicated operation.

5. $7(y+4)$ 5._____

6. $-5(z-4)$ 6._____

Simplify the following:

7. $2a-4(a-3)$ 7._____

8. $6(4y-3)-2(5y-1)$ 8._____

9. $-2.1x+9.3+5.6x+4$ 9._____

Key Terms

Use the vocabulary terms listed below to complete each statement in exercises 1–5.

    linear   equivalent   solution   equation   **Addition Property of Equality**

1. A(n) _____ is a mathematical sentence indicating that two expressions are equal.

2. A(n) _____ of an equation is any value of the variable that makes the equation true.

3. A(n) _____ equation is any equation that can be put in the form of $ax+b=c$, where a, b, and c are constants and $a \neq 0$.

4. The _____ says if the same number is added to both sides of an equation, the result is still an equation.

5. Equations that have the same solutions are called _____ equations.

Practice Problems

Objective A Solve linear equations, using the addition property of equality.

Solve each of the following.

1. $m - 4 = -9$ 1._____

2. $y + 2 = 7$ 2._____

3. $21.6 = b + 11.8$ 3._____

4. $-5p - 9 + 6p = 1$ 4._____

5. $34 - 15 + x = 7$ 5._____

6. $12q - 4 - 11q = 2 - 10$ 6._____

Name:
Instructor:

Date:
Section:

7. $-5+7=-2x+8+3x$

7._____

8. $9.7x+21-8.7x=5-9.6$

8._____

9. $2(x+3)-x=8$

9._____

10. $-5(2-x)-4x-1=3$

10._____

11. $10+2(x-4)+1-x=18$

11._____

12. $16x-5(3x+4)=9$

12._____

13. $6(x+4)-5(x+4)=9$

13._____

14. $4(2x+3)-7(x-5)=9+2$

14._____

15. $-6(7-x)+5(10-x)-8=0$

15._____

16. $3(7x+2)-4(5x+6)=19$

16._____

Objective B Translate mathematics expressions into English and English expressions into mathematics.

Translate the following English sentences into mathematical sentences and solve. Use x as the variable.

17. Some number decreased by eight is seventeen.

17._____

Copyright © 2013 Pearson Education, Inc.

18. The sum of a number and twelve is twenty-one.   18._____

19. Fourteen more than some number is thirty-two.   19._____

20. Some number minus three is equal to negative nine.   20._____

21. Five less than some number is negative eight.   21._____

22. The difference of a number and twenty is fifty.   22._____

Objective C Solve various types of real-world problems.

Let x represent the unknown in each of the following. Write an equation and solve.

23. On Monday, Darlene jogged 3 miles more than Wendy jogged. If Darlene jogged 9 miles, how far did Wendy jog?   23._____

24. Wilma needs $18,000 for a down payment on a house. If she has already saved $7400, how much more does she need?   24._____

25. Carol and Ron pooled their money to buy an anniversary gift for their parents. Ron paid $80 less than Carol. If Carol paid $135, how much did Ron pay?   25._____

26. The selling price of a refrigerator is $1875. If the markup is $565, find the store's cost.   26._____

27. Julia went on a diet and lost 32 pounds. If she weighed 146 pounds after the diet, what did she weigh before the diet?   27._____

28. The perimeter of a triangle is 29 inches. If two of the sides have lengths of 11 inches and 6 inches, what is the length of the third side?   28._____

Name:
Instructor:
Date:
Section:

Chapter 3 LINEAR EQUATIONS AND INEQUALITIES

3.2 Multiplication Property of Equality

Learning Objectives
A Solve equations by using the multiplication property of equality.
B Solve equations that require simplifications by using the multiplication property of equality.
C Solve real-world problems.

Getting Ready

Multiply or divide as indicated.

1. $5(-6)$

1._____

2. $(-12)(-4)$

2._____

3. $64 \div (-8)$

3._____

4. $-96 \div (-6)$

4._____

5. $(-.4)(2.3)$

5._____

6. $4.5 \div (-.9)$

6._____

Multiply.

7. $\dfrac{4}{3} \cdot 39$

7._____

8. $-6 \cdot \dfrac{1}{-6}$

8._____

9. $\dfrac{5}{8} \cdot \dfrac{8}{5}$

9._____

Key Terms

Use the vocabulary terms listed below to complete each statement in exercises 1–4.

> reciprocal   one   product   Multiplication Property of Equality

1. The multiplicative inverse is also called the _____.

2. The word _____ indicates multiplication.

3. The _____ says both sides of an equation may be multiplied (or divided) by any nonzero number and the result will still be an equation.

4. Two numbers are multiplicative inverses if their product is _____.

Practice Problems

Objective A Solve equations by using the multiplication property of equality.

Solve the following.

1. $7x = -14$

    1._____

2. $-9x = -81$

    2._____

3. $4b = -76$

    3._____

4. $.6x = 3.6$

    4._____

5. $\dfrac{n}{3} = 15$

    5._____

6. $-\dfrac{1}{22}q = 3$

    6._____

Name:
Instructor:
Date:
Section:

7. $-8 = -\dfrac{1}{4}m$
7._____

8. $\dfrac{1}{3}t = -8$
8._____

9. $\dfrac{3}{8}c = -\dfrac{4}{5}$
9._____

10. $\dfrac{y}{4} = 9$
10._____

11. $9a = 54$
11._____

12. $3a = 2.1$
12._____

Objective B Solve equations that require simplifications by using the multiplication property of equality.

Solve the following.

13. $7y = 30 + 5$
13._____

14. $-2z = -28 + 50$
14._____

15. $8m - 4m + 7 = 35$
15._____

16. $4x + 5x = 81$
16._____

17. $2x - 10x = -32$ 17._____

18. $2.1x + 3.4x = 110$ 18._____

19. $8w = -58 - 6$ 19._____

20. $10 + 5 = -\dfrac{3}{4}y$ 20._____

Objective C Solve real-world problems.

Translate the following mathematical sentences into English sentences in at least two ways.

21. $5x = 35$ 21._____

22. $6t = 41$ 22._____

23. $-17 = 3m$ 23._____

24. $-2.4 = 5.6y$ 24._____

Translate from English to mathematical sentences and solve. Use x as the variable.

25. Seven times a number is forty-two.   25._____

Name: Date:
Instructor: Section:

26. Three-fourths of a number is twenty-four. 26._____

27. Some number divided by negative six is negative thirty. 27._____

28. The product of five and some number is negative thirty-five. 28._____

29. The quotient of a number and nine is sixty-three. 29._____

Solve each of the following.

30. Mary has a rug that is 14 feet long and has an area of 98 square feet. How wide is the rug? 30._____

31. Andrea worked 14 hours last week and earned $115.50. What is her hourly wage? 31._____

32. Kirby works as a salesperson and is paid a 4% commission on his sales. If he receives a $1600 commission on a sale, what was the amount of the sale? 32._____

33. Jacob is cutting wire for a project. How many 2.5-inch-long pieces can he cut from a roll of wire containing 60 inches?

33._____

34. Jerome has four times as many dollars as Theo. If Jerome has $616, how much money does Theo have?

34._____

Name:
Instructor:

Date:
Section:

Chapter 3 LINEAR EQUATIONS AND INEQUALITIES

3.3 Combining Properties in Solving Linear Equations

Learning Objectives
A Solve linear equations, using both the addition and multiplication properties of equality.
B Determine whether an equation is an identity or a contradiction.
C Translate mathematical equations into English and English sentences into mathematical equations.
D Solve application problems.

Getting Ready

Add or subtract the following integers:

1. $-9-16$ 1._____

2. $32-6$ 2._____

Add or subtract the following like terms:

3. $3a-5-5a-12$ 3._____

4. $6x+12-4x$ 4._____

Add, subtract, multiply, or divide the following decimals:

5. $.7-4.1$ 5._____

6. $18(.4)$ 6._____

Multiply or divide the following integers:

7. $-2(8)$ 7._____

8. $\dfrac{-38}{2}$ 8._____

Simplify the following, using the distributive property:

9. $8(x-3)+9$ 9._____

10. $4(5x+1)-4(x-3)$ 10._____

Solve the following, using the addition property of equality:

11. $y-7=9$ 11._____

12. $3(a-2)-a+5=15$ 12._____

Solve the following, using the multiplication property of equality:

13. $4y=-2.8$ 13._____

14. $\dfrac{x}{6}=-8$ 14._____

Key Terms

Use the vocabulary terms listed below to complete each statement in exercises 1–2.

contradiction **identity**

1. A(n) _____ is an equation that is true for all values of the variable for which the equation is defined.

2. A(n) _____ is an equation that has no solution.

Name:
Instructor:

Date:
Section:

Practice Problems

Objective A   **Solve linear equations, using both the addition and multiplication properties of equality.**

Solve the following.

1. $9x - 1 = -19$ 1._____

2. $10y + 14 = 114$ 2._____

3. $8w + 6 = 38$ 3._____

4. $30z = 5z + 5$ 4._____

5. $7w + 6 = -9 + 5w$ 5._____

6. $4a + 8 = -3 - 9a$ 6._____

7. $-7m - 7 = 1 + 3m$ 7._____

8. $4x + 1 = -1 - 3x - 4x$ 8._____

9. $7y - (3y - 1) = 21$ 9._____

Copyright © 2013 Pearson Education, Inc.

10. $4w+5-5w=6w+12$ 10._____

11. $2.6x+1.3=-.4x-1.7$ 11._____

12. $7(w-6)=4w+3$ 12._____

13. $5(b+1)-7(2b-6)+5=79$ 13._____

14. $-3(n-4)+2(6-n)=49$ 14._____

15. $10(2+3n)-22n=6(n-1)$ 15._____

16. $6b-3-9(1+b)=3(5b-4)$ 16._____

17. $-8=\dfrac{x+3}{7}$ 17._____

18. $13-3t=26+6(1-t)$ 18._____

Objective B Determine whether an equation is an identity or a contradiction.

Determine whether the following are identities or contradictions and indicate the solutions.

19. $4(x+3)-(4x+12)=0$ 19._____

126   Copyright © 2013 Pearson Education, Inc.

Name:
Instructor:

Date:
Section:

20. $12(3y-10)=6(6y+4)$ 20._____

21. $(x-3)-(x+7)=4$ 21._____

22. $2q+5(q-4)=4q+3(q-8)+4$ 22._____

Objective C Translate mathematical equations into English and English sentences into mathematical equations.

Translate the following English sentences into mathematical sentences and solve. Use x as the variable.

23. Six less than three times a number is fifteen. 23._____

24. Seven times a number less the difference of three times the number and one is equal to two. 24._____

25. Four times a number minus seven is twenty-nine. 25._____

26. One hundred fourteen is the same as fourteen more than the product of ten and the number. 26._____

Objective D Solve application problems.

Solve the following.

27. A rectangular carpet has a perimeter of 264 inches. The length of the carpet is 12 inches more than the width. What are the dimensions of the carpet?

27._____

28. A square stage has a perimeter which is 11 times the length of a side, decreased by 21. Find the length of a side.

28._____

29. An electrician charges $70 for a service call and $65 per hour base rate. His charge for a repair job was $421, excluding parts. How many hours did the electrician work?

29._____

30. Saline solution has 1 part salt to 8 parts water by weight. Find the number of ounces of water and the number of ounces of salt in 171 ounces of the solution.

30._____

Name: Date:
Instructor: Section:

Chapter 3 LINEAR EQUATIONS AND INEQUALITIES

3.4 Using and Solving Formulas

Learning Objectives
A Solve for the unknown when given all of the values in a formula except one.
B Solve a formula for a given variable.

Getting Ready

Solve the following equations, using the addition and multiplication properties of equality:

1. $42 = 18 + 8a$ 1._____

2. $45 = \dfrac{9y}{3}$ 2._____

3. $10.764 = 4.14x$ 3._____

Key Terms

Use the vocabulary terms listed below to complete each statement in exercises 1–2.

        formula   **variable**

1. A formula can be solved for any_____ in terms of the other variables of that formula.

2. A _____ is an equation that expresses a relationship that applies to a whole class of problems.

Practice Problems

Objective A Solve for the unknown when given all of the values in a formula except one.

Solve each formula for the indicated variable with the given information.

1. $A = \dfrac{bh}{2}$ for A with $b = 5$ inches and $h = 10$ inches 1._____

Copyright © 2013 Pearson Education, Inc.

2. $d = rt$ for t with $d = 125$ miles and $r = 50$ miles per hour

   2._____

3. $C = 2\pi r$ for r with $C = 75.36$ feet and $\pi = 3.14$

   3._____

4. $P = 2W + 2L$ for L with $W = 18$ meters and $P = 88$ meters

   4._____

5. $I = PRT$ for P with $I = 1008$, $R = .035$, and $T = 9$

   5._____

6. $P = a + b + c$ for a with $P = 62$, $b = 12$, and $c = 21$

   6._____

7. $V = IR$ for I with $V = 120$ and $R = 25$

   7._____

8. $P = \dfrac{F}{A}$ for F with $P = 60$ and $A = 10$

   8._____

9. $R = \dfrac{I}{PT}$ for T with $R = .06$, $I = 405$, and $P = 1500$

   9._____

10. $2x - 5y = 18$ for x with $y = -\dfrac{4}{5}$

    10._____

Objective B   Solve a formula for a given variable.

Solve each formula for the indicated variable.

11. $A = \dfrac{h}{2}(b + B)$ for h

    11._____

130   Copyright © 2013 Pearson Education, Inc.

Name:
Instructor:

Date:
Section:

12. $V = LWH$ for H

12._____

13. $P = IV$ for I

13._____

14. $C = 2\pi r$ for r

14._____

15. $A = \dfrac{1}{2}bh$ for h

15._____

16. $S = 4LW + 2WH$ for L

16._____

17. $6x + 8y = 7$ for y

17._____

18. $-9x + 2y = 12$ for x

18._____

19. $A = P + Prt$ for t

19._____

20. $A = \pi r^2$ for r^2

20._____

Name: Date:
Instructor: Section:

Chapter 3 LINEAR EQUATIONS AND INEQUALITIES

3.5 General, Consecutive Integer, and Distance Application Problems

Learning Objectives
A Solve real-world problems that are general in nature.
B Solve problems involving consecutive, consecutive odd, and consecutive even integers.
C Solve real-world problems involving distance, time, and rate.

Getting Ready

Solve by using both the addition and multiplication properties of equality.

1. $4x + 14{,}864 = 271{,}492$ 1._____

2. $215 + .5x = 475$ 2._____

3. $x + (x+2) + (x+4) = 57$ 3._____

4. $60x = 8(x-5)$ 4._____

Key Terms

Use the vocabulary terms listed below to complete each statement in exercises 1–4.

  consecutive odd integers   time   consecutive integers   consecutive even integers

1. _____ are integers that follow one another in order.

2. _____ are every other integer, starting with an even integer.

3. _____ are every other integer, starting with an odd integer.

4. Distance equals rate times _____.

Copyright © 2013 Pearson Education, Inc.   133

Practice Problems

Objective A Solve real-world problems that are general in nature.

Solve the following general application problems.

1. A high school graduating class is made up of 1480 students. There are 176 more girls than boys. How many boys are in the class?

   1._____

2. Walt's job pays him $80 per day plus 4% commission on all his sales. If his total wages on Wednesday were $152, find the amount of his sales.

   2._____

3. In a local election William Barnes received 1080 more votes than Thomas Williamson. If a total of 11,640 votes were cast, how many did Thomas Williamson receive?

   3._____

4. Elsa makes three times as much money as Helen. If the total of their salaries is $250,000, find Elsa's salary.

   4._____

Name: Date:
Instructor: Section:

5. If the sum of a number and seven is doubled, the result is eighteen less than triple the number. Find the number.

5._____

6. A car rental agency advertised their rate for an economy car as $35.95 per day and $0.28 per mile. If you rent this car for 3 days, how many whole miles can you drive on a $248 budget?

6._____

Objective B   Solve problems involving consecutive, consecutive odd, and consecutive even integers.

Solve the following consecutive integer problems.

7. The sum of two consecutive integers is 2471. Find the integers.

7._____

8. Find two consecutive even integers such that five times the larger integer is sixty-four more than twice the smaller integer.

8._____

9. The sum of three consecutive integers is 369. Find the integers.

9._____

10. If the first and third of three consecutive odd integers are added, the result is thirty-four less than the middle integer. Find the integers.

10._____

11. If three times the smaller of two consecutive integers is added to four times the larger, the result is 123. Find the larger integer.

11._____

12. Two consecutive pages in a book have 401 as the sum of their page numbers. What is the number of the second page?

12._____

13. Find two consecutive integers such that six times the larger is twenty-five less than seven times the smaller.

13._____

Name: Date:
Instructor: Section:

14. The score at the football game between Evans   14._____
    Community College and Southern State College
    was two consecutive even integers with Southern
    State winning. If the sum of the scores was 58,
    what was Southern State's winning score?

Objective C   Solve real-world problems involving distance, time, and rate.

Solve the following distance, rate, and time problems.

15. A car traveling 65 miles per hour overtakes a truck   15._____
    traveling 40 miles per hour that had a one-hour
    head start when they left Houston. How far from
    Houston are the vehicles?

16. Two joggers are 8 miles apart and are jogging   16._____
    toward each other. They meet in 1 hour. Find the
    rate of each jogger if one jogs .8 mile per hour
    faster than the other.

17. Two trains leave cities 380 miles apart at 8:00 a.m.   17._____
    traveling toward each other. One train travels at 50
    mph and the other train travels at 45 mph. At what
    time do they pass each other?

Copyright © 2013 Pearson Education, Inc.   137

18. Scott and Dana leave their home traveling in opposite directions. Scott drives 20 mph faster than Dana. How fast is Dana driving if they are 300 miles apart after 3 hours?

18._____

19. Carolyn can drive from Rocky Mount to Staunton at a certain speed in 6 hours. If she increases her speed by 10 mph she can make the trip in 5 hours. How far is Rocky Mount from Staunton?

19._____

20. Thelma averaged 50 mph on a recent trip to Springfield. On the return trip, traffic was heavier so she took a detour. The return trip took 2 hours longer and was 30 miles longer. If Thelma averaged 42 mph on the return trip how many hours did the return trip take?

20._____

Name:
Instructor:

Date:
Section:

Chapter 3 LINEAR EQUATIONS AND INEQUALITIES

3.6 Money, Investment, and Mixture Application Problems

Learning Objectives
A   Solve real-world problems involving money.
B   Solve real-world problems involving investments.
C   Solve real-world problems involving mixtures.

Getting Ready

Solve, using both the addition and multiplication properties of equality.

1. $8x + 6(12 - x) = 80$ 1._____

2. $2.6x + 17.4 = 4.2(x + 3)$ 2._____

3. $.8x + 3.95 = .3(x + 16)$ 3._____

Key Terms
Use the vocabulary terms listed below to complete each statement in exercises 1–4.

   mixture problems   **500 + x**   **500 − x**   **investment problems**

1. _____ are special types of money problems.

2. Investment problems and _____ are solved basically the same way.

3. If 500 is split into two parts and one of the parts is x, then the other part is _____.

4. If some quantity is 500 more than an amount x, then it can be represented by _____.

Practice Problems

Objective A Solve real-world problems involving money.

Solve the following money problems.

1. Bobby paid $5250 for 28 doors. If the inside doors cost $180 each and the outside doors cost $210 each, how many of each kind did he buy?

    1._____

2. Pamela spent $20.68 for 60 stamps. If some of the stamps cost 44¢ each and the remainder cost 18¢ each, how many of each kind did she buy?

    2._____

3. Cassie went to the office supply store and spent $106 on 38 packs of pencils and pens. The packs of pencils cost $1.25 each while the packs of pens were $4.50 each. How many of each kind did she buy?

    3._____

Name: Date:
Instructor: Section:

4. Geraldine has been saving dimes and quarters in a jar. She decided to count her money and discovered that she had $47. If she has 20 more quarters than dimes, how many of each type coin does she have?

4._____

5. Teri set up a concession stand at the baseball game. She sold bottles of water for $1.25 each and bags of popcorn for $2 each. At the end of the day she had collected $190 and had sold a total of 125 items. How many bags of popcorn did she sell?

5._____

Objective B Solve real-world problems involving investments.

Solve the following investment problems.

6. William invested $5000 in two different accounts. The first account was very safe and paid 3% annual interest while the second account was risky but it paid 18% annual interest. If he earned $375 from the two accounts for the year, how much money did he invest in each account?

6._____

7. Neo invested some money at 6% annual interest and $800 more than that amount at 7.5% annual interest. If his total yearly interest was $181.50, how much was invested at each rate?

7._____

8. Wilma invests $12,000 at 5% annual interest. How much must she invest at 8% annual interest if the total interest earned per year is $1800?

8._____

9. Keanu invested some money in an account that was paying 6.5% annual interest. He invested triple that amount in an account that paid 9% annual interest. At the end of the year he had earned $268 in interest. How much did he invest in the account paying 9%?

9._____

Name: Date:
Instructor: Section:

Objective C Solve real-world problems involving mixtures.

Solve the following mixture problems.

10. How much of an alloy that is 20% copper should be mixed with 500 ounces of alloy that is 45% copper in order to get an alloy that is 30% copper?

    10._____

11. How much water should be added to 30 gallons of a solution that is 60% antifreeze in order to get a mixture that is 40% antifreeze?

    11._____

12. A chemist wants to make 110 ml of a 30% hydrochloric acid solution. The only solutions she has in stock are 10% and 60% hydrochloric acid. How much of each must she use?

    12._____

13. Sherri is putting together 40 pounds of trail mix made up of dried fruit and nuts. She plans on selling the mix for $3 per pound. How many pounds of each should she use if the dried fruit sells for $2.50 per pound and the nuts sell for $5 per pound?

13. _____

14. Beans Galore is preparing a blend of coffee that costs $9.55 per pound from coffee beans that cost $6.25 per pound and others that cost $11.75 per pound. How many pounds of each type bean must be used to prepare 50 pounds of this special blend?

14. _____

Name: Date:
Instructor: Section:

Chapter 3 LINEAR EQUATIONS AND INEQUALITIES

3.7 Geometric Application Problems

Learning Objectives
A   Solve real-world problems involving geometric concepts.

Getting Ready

Solve, using both the addition and multiplication properties of equality.

1. $50 = 4(3x+2) - 5x$ 1._____

2. $-37 = 2y + (y-5) + (8y+1)$ 2._____

3. $z + 5z + (z+17) = 80$ 3._____

Key Terms

Use the vocabulary terms listed below to complete each statement in exercises 1–7.

transversal	supplementary	parallel	congruent
degrees	complementary	triangle	

1. Two angles are _____ if the sum of their measures is 180°.

2. Two angles are _____ if the sum of their measures is 90°.

3. The size of an angle is measured in _____.

4. The sum of the measures of all the angles of any _____ is 180°.

5. Two or more lines are _____ if they lie in the same plane and do not intersect.

6. Any line intersecting two or more lines is called a(n) _____.

7. Two angles with the same measure are called _____.

Copyright © 2013 Pearson Education, Inc.   145

Practice Problems

Objective A Solve real-world problems involving geometric concepts.

Solve the following geometric problems.

1. If the length of a rectangular lot is 40 feet less than triple its width, and the perimeter is 160 feet, find the length of the lot.

    1._____

2. A rectangular conference room is 20 feet longer than it is wide. If the perimeter is 168 feet, what are its length and width?

    2._____

3. A triangular shaped flower bed has one side four times the length of the shortest side and the third side is 25 feet longer than the shortest side. Find the dimensions if the perimeter is 61 feet.

    3._____

4. Two sides of a triangular lot are equal. The third side is 6 feet less than the other two sides. If the perimeter is 54 feet. Find the lengths of the three sides.

    4._____

Name: Date:
Instructor: Section:

5. The measures of the two smaller angles in a triangle are the same and the third angle is double the measure of the smaller angle. Find the measure of each angle.

5._____

6. An angle measures six less than three times the measure of its complement. Find the measure of each angle.

6._____

7. An angle measures four times the measure of its supplement. Find the measure of each angle.

7._____

8. Kara has a triangular sign where two of the angles have the same measure. If the measure of the third angle is 15° more than each of the other angles, what are the measures of the angles?

8._____

9. A triangular piece of property has a perimeter of 1800 feet. One side is 100 feet longer than the shortest side, while the remaining side is 200 feet longer than the shortest side. Find the lengths of all three sides.

9._____

10. An angle is 27° less than twice its complement. Find the angle.

10._____

11. How large is an angle if it is 12° more than triple its supplement?

11._____

12. The measure of the smallest angle of a triangle is .4 of the measure of the largest angle, and the measure of the middle angle is .6 of the measure of the largest angle. Find all three angles.

12._____

Name:
Instructor:

Date:
Section:

13. One angle of a triangle measures 90°. One of the remaining two angles is six less than five times the measure of the other remaining angle. Find the measure of the remaining two angles.

13._____

14. Given the following figure, where L_1 and L_2 are parallel, find the value of x and $\angle A$ and $\angle B$.

14._____

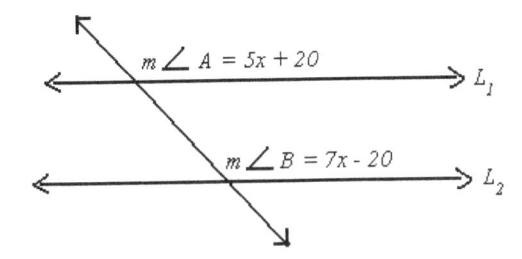

$m\angle A = 5x + 20$ → L_1

$m\angle B = 7x - 20$ → L_2

15. Given the following figure, where L_1 and L_2 are parallel, find the value of x and $\angle A$ and $\angle B$.

15._____

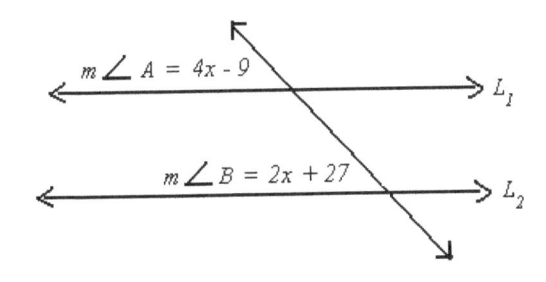

$m\angle A = 4x - 9$ → L_1

$m\angle B = 2x + 27$ → L_2

Copyright © 2013 Pearson Education, Inc.

16. Given the following figure, where L_1 and L_2 are parallel, find the value of x and $\angle A$ and $\angle B$.

16._____

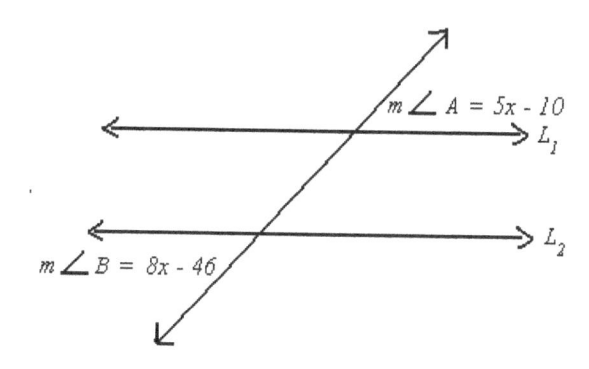

$m\angle A = 5x - 10$
$m\angle B = 8x - 46$

17. Given the following figure, where L_1 and L_2 are parallel, find the value of x and $\angle A$ and $\angle B$.

17._____

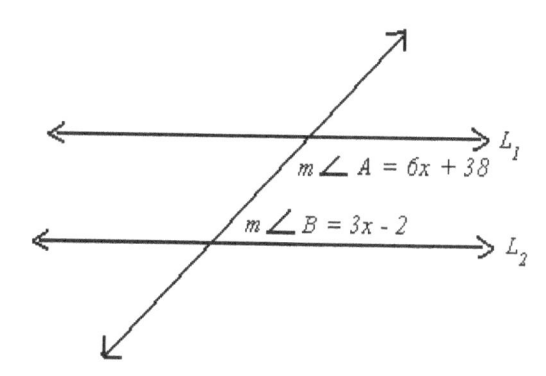

$m\angle A = 6x + 38$
$m\angle B = 3x - 2$

18. Given the following figure, where L_1 and L_2 are parallel, find the value of x and the measures of $\angle A$ and $\angle B$.

18._____

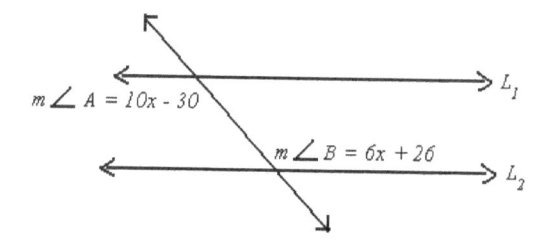

$m\angle A = 10x - 30$
$m\angle B = 6x + 26$

Name:
Instructor:
Date:
Section:

Chapter 3 LINEAR EQUATIONS AND INEQUALITIES

3.8 Solving Linear Inequalities

Learning Objectives
- A Graph inequalities on the number line.
- B Solve linear inequalities, using the addition property of inequality, and graph the solutions.
- C Solve linear inequalities, using the multiplication property of inequality, and graph the solutions.
- D Solve linear inequalities, using both the addition and multiplication properties of inequality, and graph the solutions.
- E Solve compound inequalities.
- F Translate English sentences into mathematical inequalities and vice versa.

Getting Ready

Solve the following equations, using the addition property of equality:

1. $a - 7 = 8$

    1._____

2. $3(5b - 3) - 7(2b + 4) = -30$

    2._____

Solve the following equations, using the multiplication property of equality:

3. $-5x = 20$

    3._____

4. $\dfrac{y}{-4} = 8$

    4._____

Solve the following equations, using both the addition and multiplication properties of equality:

5. $5x - 4 = x - 7$

    5._____

6. $-4(x + 2) + 5x = 3(x + 4) + x$

    6._____

Key Terms

Use the vocabulary terms listed below to complete each statement in exercises 1–7.

inequality   compound   open dot   direction
irrational   dot   Addition Property of Inequality

1. The _____ of an inequality changes when we multiply or divide both sides of the inequality by a negative number.

2. The numbers between the integers on the number line are the rational and _____ numbers.

3. $x > 2$ is an example of a(n) _____.

4. A(n) _____ on a number line indicates the point is a part of the solution.

5. A(n) _____ on a number line indicates the point is not a part of the solution.

6. An inequality of the form $a < x < b$ is called a(n) _____ inequality.

7. The _____ says, any real number may be added to (or subtracted from) both sides of an inequality without affecting the order.

Practice Problems

Objective A Graph inequalities on the number line.

Graph each of the following.

1. $x > 5$ 1. See graph _____

2. $x \leq -4$ 2. See graph _____

3. $x \geq -1$ 3. See graph _____

Name: Date:
Instructor: Section:

4. $x < 7$ **4.** See graph

5. $-2 < x < 3$ **5.** See graph

6. $-1 \leq x \leq 5$ **6.** See graph

7. $0 \leq x < 4$ **7.** See graph

Objective B Solve linear inequalities, using the addition property of inequality, and graph the solutions.

Solve by using the addition property of inequality and graph the solutions.

8. $x + 3 < 5$ **8.** See graph

9. $x - 2 \geq 6$ **9.** See graph

10. $6y - 2 - 2y \leq 6$ **10.** See graph

11. $8q + 3 - 5 - 7q > 0$ **11.** See graph

12. $4(m+1) - 3m \geq 7$ **12.** See graph

Copyright © 2013 Pearson Education, Inc.

13. $-8(n-4)+9n-27<4-6$ 13. See graph

←++++++++++++++++++++++++++++→

Objective C   Solve linear inequalities, using the multiplication property of inequality, and graph the solutions.

Solve by using the multiplication property of inequality.

14. $6x > -18$ 14._____

15. $-3y \leq -6$ 15._____

16. $-7a \geq 28$ 16._____

17. $\dfrac{w}{-6} < 2$ 17._____

18. $\dfrac{m}{4} \leq 1$ 18._____

19. $-8q > -1.6$ 19._____

Objective D   Solve linear inequalities, using both the addition and multiplication properties of inequality, and graph the solutions.

Solve by using the addition and multiplication properties of inequality.

20. $2p+7 < 11$ 20._____

154   Copyright © 2013 Pearson Education, Inc.

Name:
Instructor:

Date:
Section:

21. $-4z + 7 \leq 19$

21._____

22. $20x - 30 > 5(3x - 5)$

22._____

23. $-6(3b - 1) < -24b - 48$

23._____

24. $-28t + 8 \leq -4(6t - 4)$

24._____

25. $30y - 10 \leq 5(5y - 11)$

25._____

26. $-5x + 6 - 2x < 4 - 9x + 4$

26._____

Objective E   Solve compound inequalities.

Solve by using the addition and multiplication properties of inequality..

27. $-3 < x + 2 \leq 6$

27._____

28. $-6 \leq 2y+4 < 8$ 28._____

29. $-7 \leq 12y+3 \leq 10$ 29._____

30. $-8 < 4-3a < 16$ 30._____

Objective F Translate English sentences into mathematical inequalities and vice versa.

Translate the following English sentences into mathematical sentences and solve. Use x as the variable.

31. Three times some number is greater than negative twenty-four.   31._____

32. Seven times the sum of a number and six is at most fourteen.   32._____

33. Nine times the difference of a number and one is at least thirty-six.   33._____

34. Sixteen is greater than or equal to twice the difference of five and a number.   34._____

Name:
Instructor:

Date:
Section:

Chapter 4 GRAPHING LINEAR EQUATIONS AND INEQUALITIES

4.1 Reading Graphs and the Cartesian Coordinate System

Learning Objectives
A   Read and interpret bar, pie, and line graphs.
B   Verify solutions of equations with two variables.
C   Write solutions of linear equations with two variables as ordered pairs.
D   Determine whether a given ordered pair is a solution of a given equation.
E   Find the missing member in an ordered pair for a given linear equation.
F   Plot points on the rectangular coordinate system.
G   Draw and interpret scatter diagrams.

Getting Ready

Add, subtract, multiply, or divide the following integers:

1. $-6+8$ 1._____

2. $9-(-6)$ 2._____

3. $-5(8)$ 3._____

4. $-18 \div 3$ 4._____

Substitute $x=-5$ and $y=4$ into the given expressions and evaluate.

5. $2x-4y$ 5._____

6. $-6x+y$ 6._____

Solve the following equations:

7. $15-8x=-1$ 7._____

8. $4x+5=-27$ 8._____

Copyright © 2013 Pearson Education, Inc.

Key Terms

Use the vocabulary terms listed below to complete each statement in exercises 1–7.

> scatter diagram   ordered pair   circle graph   x-axis
> y-axis   quadrants   origin

1. Another name for a pie chart is _____.

2. The x- and y-axes divide the rectangular coordinate system into _____.

3. The horizontal axis in the rectangular coordinate system is called the _____.

4. The vertical axis in the rectangular coordinate system is called the _____.

5. The point where the horizontal and vertical axes intersect in the rectangular coordinate system is called the _____.

6. A(n) _____ consists of two numbers enclosed in parentheses and separated by a comma.

7. When paired data are graphed in the rectangular coordinate system, the resultant graph is called a(n) _____.

Practice Problems

Objective A   Read and interpret bar, pie, and line graphs.

Following is a pie chart showing the number of newly hired teachers in Florida in 2006.

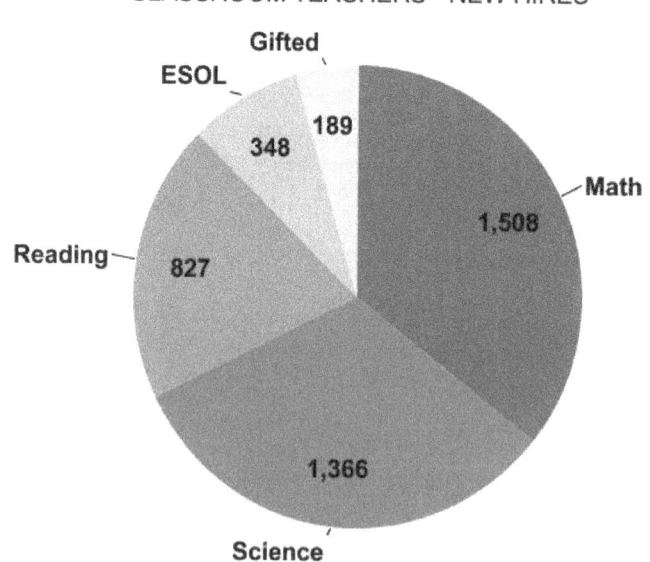

CLASSROOM TEACHERS - NEW HIRES

- Gifted: 189
- ESOL: 348
- Math: 1,508
- Reading: 827
- Science: 1,366

(Source: Florida Department of Education)

158

Name:
Instructor:

Date:
Section:

1. Which area has the largest number of new hires? 1._____

2. How many new hires were there in 2006 in the five areas listed? 2._____

3. How many more new hires are in ESOL than in Gifted? 3._____

The following bar graph shows the birth rate (number of births per 1000 population) for 8 countries.

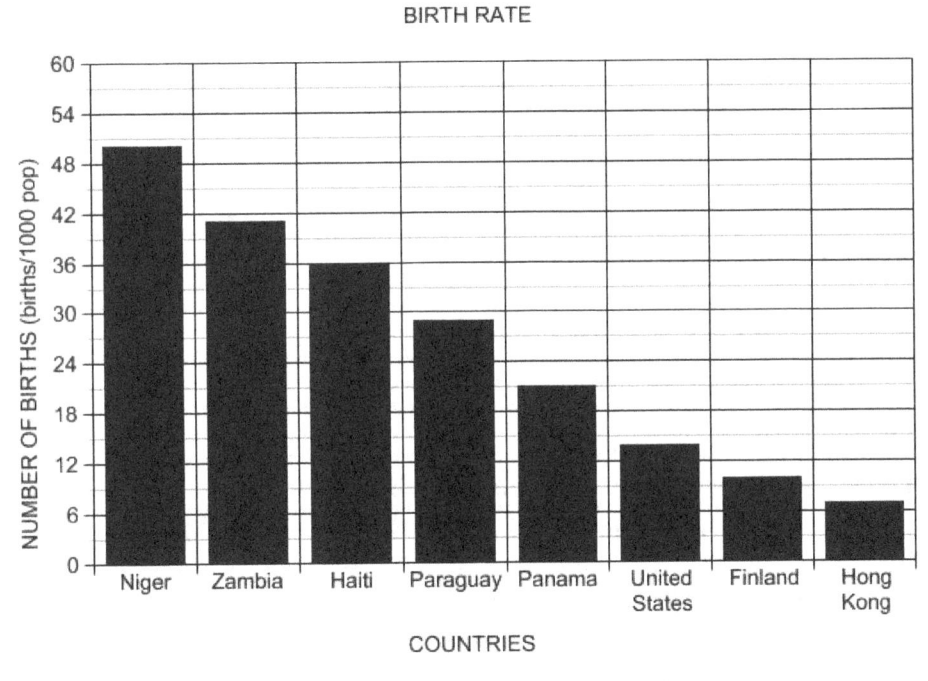

BIRTH RATE

(Source: CIA - The World Factbook 2007)

4. Which country has the lowest birth rate per 1000 population and what is that rate? 4._____

5. Which countries have a birth rate of more than 30 per 1000 population? 5._____

6. Find the difference in birth rates between Haiti and Finland. 6._____

The following line graph shows the annual sales (in millions of dollars) for U.S. full-service restaurants from 1998 to 2009.

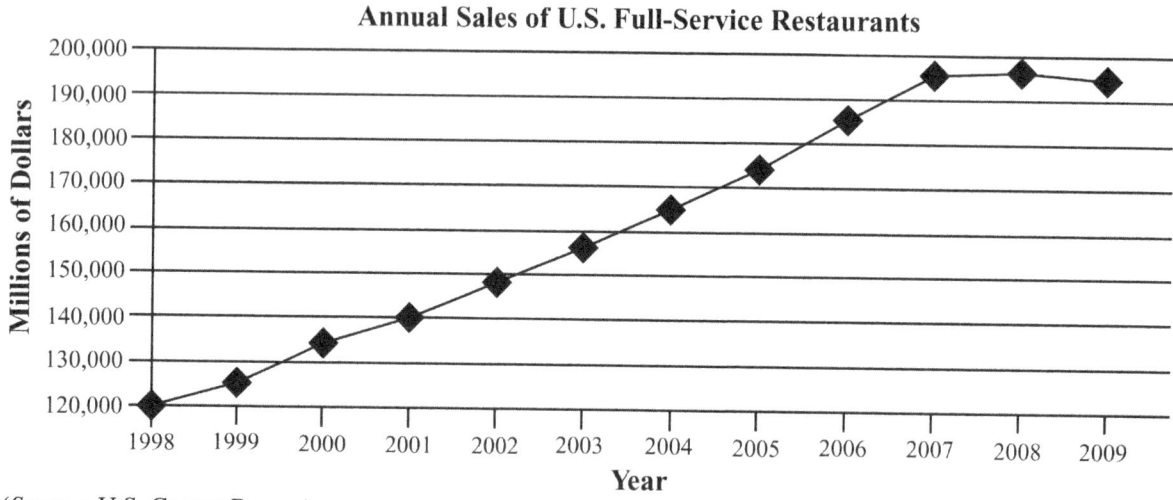

Annual Sales of U.S. Full-Service Restaurants

(Source: U.S. Census Bureau)

7. Between which two years did annual sales of full-service restaurants increase the most?

7._____

8. What was the general trend in annual sales of full-service restaurants between 2007 and 2009?

8._____

9. Estimate the annual sales for full-service restaurants in 2005.

9._____

Objectives B, C   Verify solutions of equations with two variables. Write solutions of linear equations with two variables as ordered pairs.

Verify that the following are solutions of the given equations. Then write the solutions as ordered pairs.

10. $2x + y = 8$; $x = 3$, $y = 2$

10._____

11. $x - 3y = 9$; $x = 6$, $y = -1$

11._____

Name:
Instructor:
Date:
Section:

Objective D Determine whether a given ordered pair is a solution of a given equation.

Determine whether the given ordered pair is a solution of the given equation. Assume that the ordered pairs are of the form (x, y).

12. $y = -3x + 5$; $(2, -1)$ 12._____

13. $y = 4x - 11$; $(-2, -3)$ 13._____

14. $6x - y = 19$; $(4, -5)$ 14._____

15. $7x - 8y = -17$; $(1, 3)$ 15._____

Objective E Find the missing member in an ordered pair for a given linear equation.

Use the given equation to find the missing member of each of the given ordered pairs. Assume that the ordered pairs are of the form (x, y).

16. $y = 3x - 8$; $(0, \_\_)$, $(\_\_, 0)$, $(1, \_\_)$ 16._____

17. $y = -3x + 7$; $(0, \_\_)$, $(\_\_, 0)$, $(\_\_, 1)$ 17._____

18. $3x + 8y = 24$; $(0, \_\_)$, $(\_\_, 0)$, $(-8, \_\_)$ 18._____

19. $-4x+5y=20$; $(0,\_\_)$, $(\_\_,0)$, $(\_\_,-4)$

19._____

Objective F Plot points on the rectangular coordinate system.

Plot the points represented by the given ordered pairs on the rectangular coordinate system. Assume that each unit represents 1.

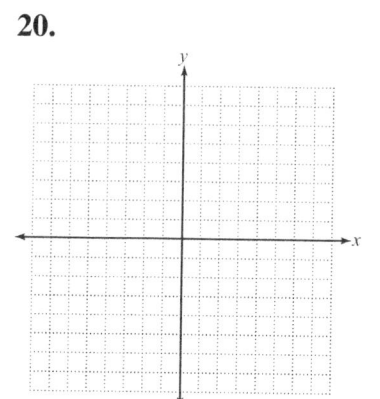

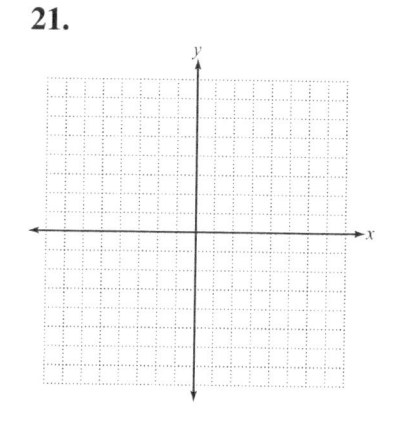

20. $(2,0)$, $(-4,1)$, $(-3,5)$, $(6,-2)$

20.

21. $(-1,0)$, $(-5,0)$, $(0,6)$, $(3,2)$

21.

Name:
Instructor:

Date:
Section:

22. $(-2,-4)$, $(0,4)$, $(-3,-5)$, $(7,0)$

22.

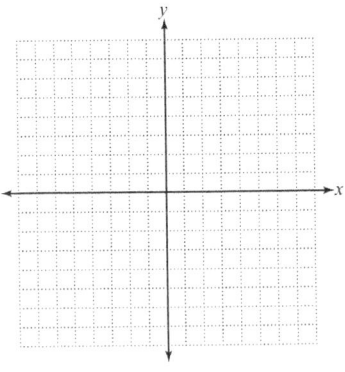

Objective G Draw and interpret scatter diagrams.

Following is a table showing the total housing units sold in Loudoun County, Virginia in 2006.

Month	Units
January	377
February	387
March	552
April	467
May	484
June	515
July	425
August	448
September	383
October	372
November	362
December	472

23. Write this paired data as ordered pairs in the form of (month, units sold).

23._____

24. What does the ordered pair (June, 515) represent?

24._____

Copyright © 2013 Pearson Education, Inc. 163

25. Use the ordered pairs to draw a scatter diagram.   **25.** ___See graph___

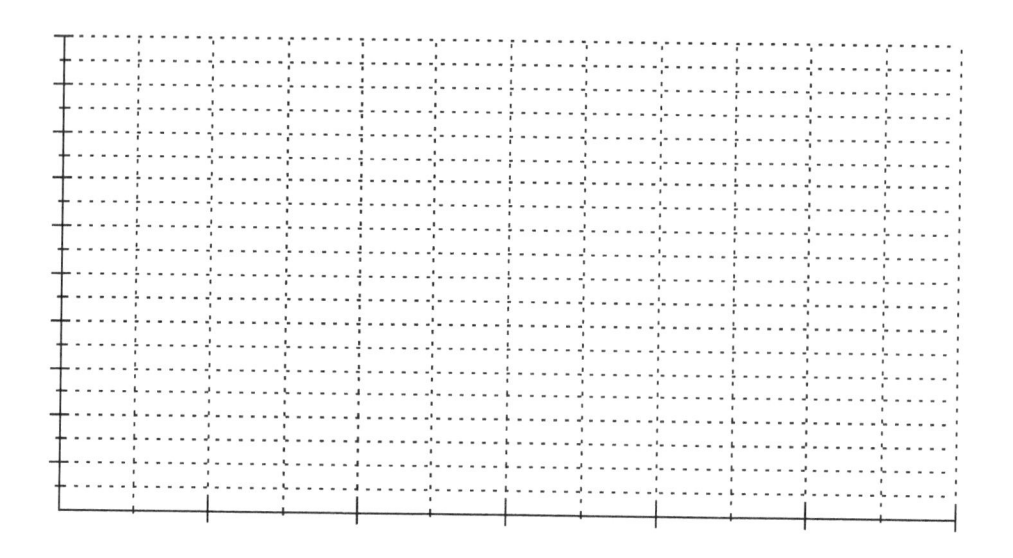

26. What was the general trend in sales in Loudoun County in 2006?   **26.** _____

164   Copyright © 2013 Pearson Education, Inc.

Name: Date:
Instructor: Section:

Chapter 4 GRAPHING LINEAR EQUATIONS AND INEQUALITIES

4.2 Graphing Linear Equations with Two Variables

Learning Objectives
A   Graph linear equations with two variables on the rectangular coordinate system.

Getting Ready

Add, subtract, multiply, or divide the following integers:

1. $-8-3$ 1._____

2. $-12-(-4)$ 2._____

3. $-4(-6)$ 3._____

4. $-27 \div (-9)$ 4._____

Substitute the given value(s) for the variable(s) into the given expressions and evaluate:

5. $-7x+2: \ x=-2$ 5._____

6. $3x-5y: \ x=6, \ y=-7$ 6._____

Solve the following equations:

7. $-4-3x=-19$ 7._____

8. $0=5x+15$ 8._____

Plot the following ordered pairs on the rectangular coordinate system:

9. $(0,-4)$, $(3,-2)$, $(-4,1)$

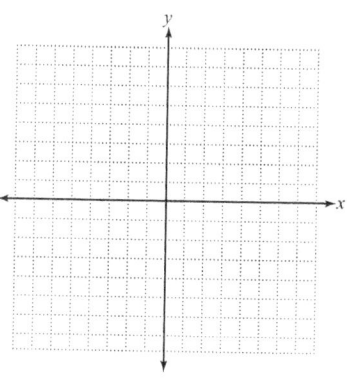

9.

Key Terms

Use the vocabulary terms listed below to complete each statement in exercises 1–5.

graph　　**horizontal line**　　**straight line**　　**vertical line**　　**linear equation**

1. A _____ is determined by two points.

2. The standard form of a _____ is $ax + by = c$, where $a, b,$ and c are integers and $a > 0$.

3. For any constant k, the graph of an equation of the form $x = k$ is a _____ line.

4. The plot of the solutions of equations with two variables is called the _____ of the equation.

5. For any constant k, the graph of an equation of the form $y = k$ is a _____ line.

Name: Date:
Instructor: Section:

Practice Problems

Objective A Graph linear equations with two variables on the rectangular coordinate system.

Complete the ordered pairs and use them to draw the graph of each equation.

1. $y = 2x + 1$; $(0, \_\_)$, $(-1, \_\_)$, $(1, \_\_)$

    1.

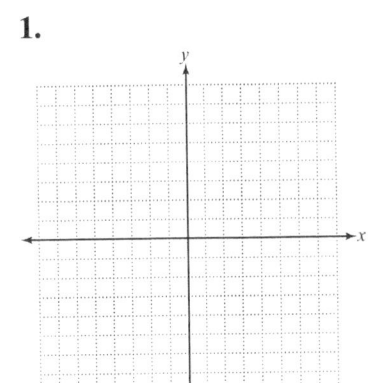

2. $x - y = 4$; $(0, \_\_)$, $(\_\_, 0)$, $(1, \_\_)$

    2.

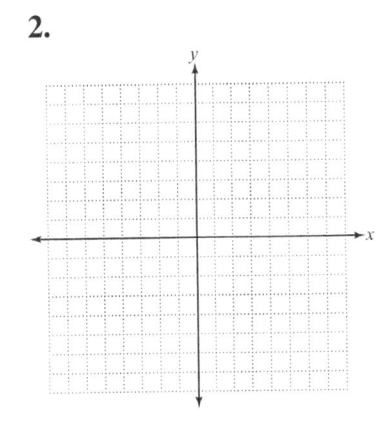

3. $3x + y = 5$; $(0, \_\_)$, $(\_\_, 0)$, $(1, \_\_)$

    3.

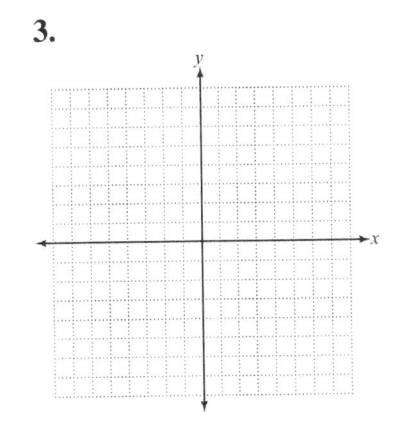

Find three ordered pairs that solve the equation and draw the graph of each.

4. $y = x + 5$

5. $y = -x + 6$

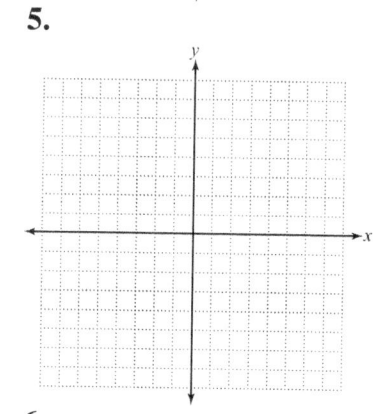

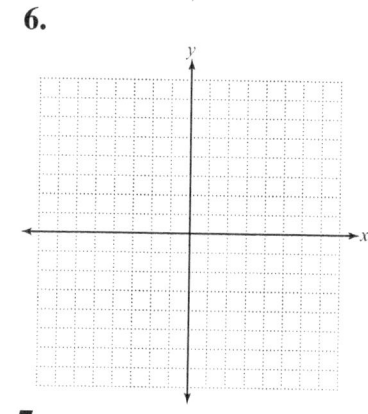

6. $y = -2x + 7$

7. $y = 3x - 2$

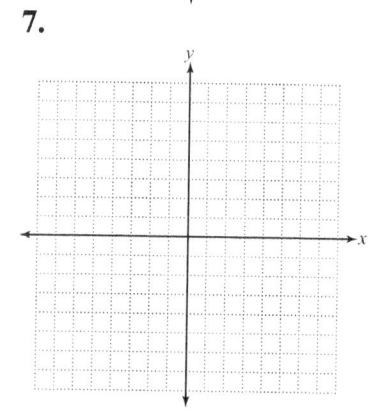

Name:
Instructor:

Date:
Section:

8. $x + y = 7$

8.

9. $x - y = 8$

9.

10. $2x + 3y = 6$

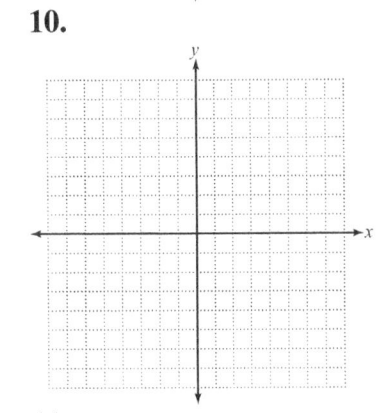

10.

11. $5x - 7y = 35$

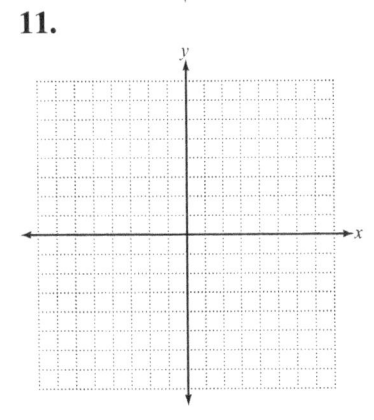

11.

Copyright © 2013 Pearson Education, Inc. **169**

12. $2x + 3y = 9$

12.

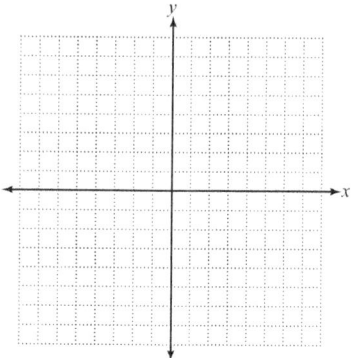

13. $7x - 2y = 6$

13.

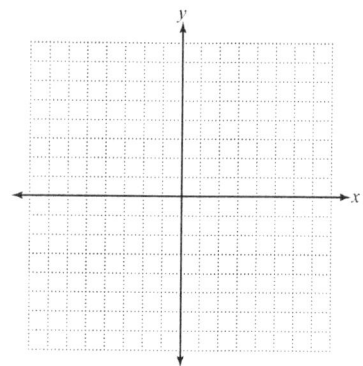

14. $-3x + 4y = 8$

14.

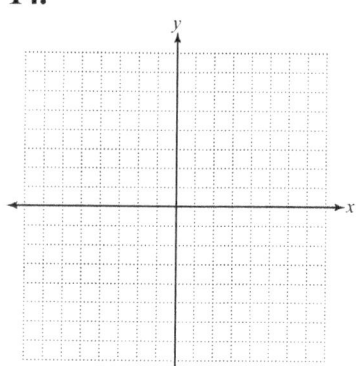

15. $-2x + 5y = -10$

15.

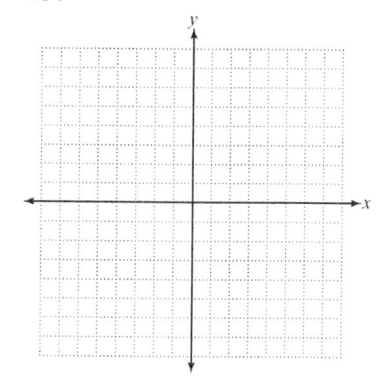

Name:
Instructor:

Date:
Section:

16. $y = -4x$

16.

17. $y = 7x$

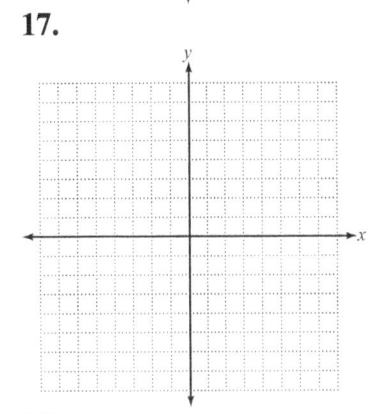

17.

18. $2x - y = 0$

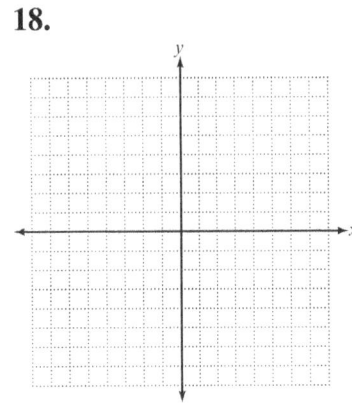

18.

19. $5x + y = 0$

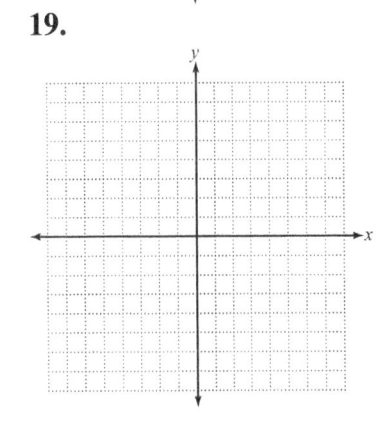

19.

Copyright © 2013 Pearson Education, Inc.   171

20. $y = \dfrac{2}{5}x$

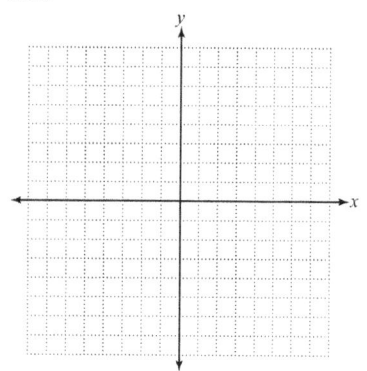

21. $y = -\dfrac{1}{4}x$

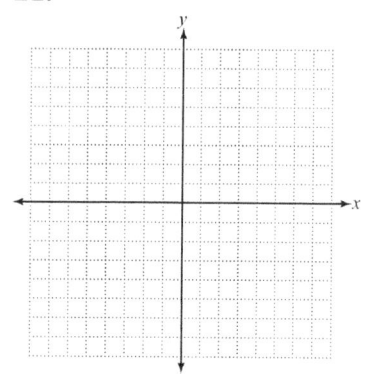

22. $x = 4$

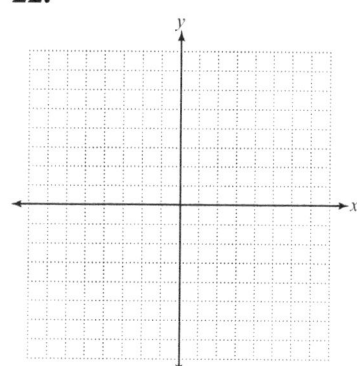

23. $y = -3$

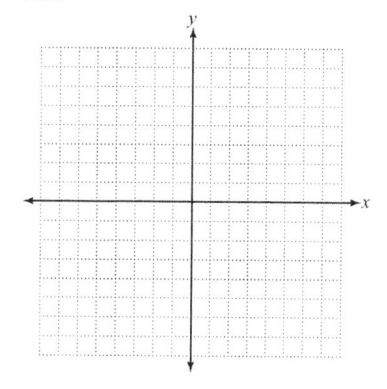

Name:
Instructor:

Date:
Section:

24. $x + 6 = 0$

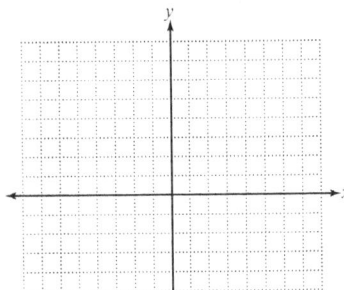

24.

25. $y - 5 = 0$

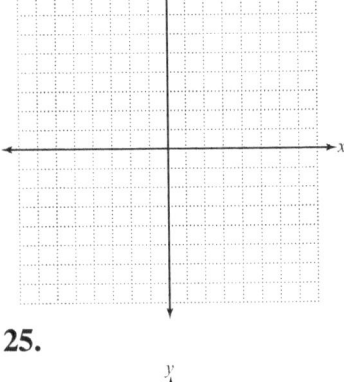

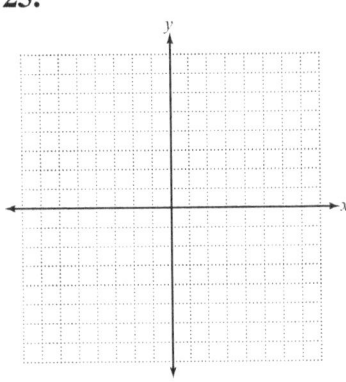

25.

The cost for renting a truck is $30 per day plus $0.20 per mile. If x represents the number of miles and y represents the cost in dollars, then $y = .2x + 30$.

26. Find the cost of a one-day rental if the truck was driven 60 miles.

26. _____

27. If the cost is $44 for a one-day rental, how many miles was the truck driven?

27. _____

28. Draw the graph if the number of miles is less than 200.

28. ___See graph___

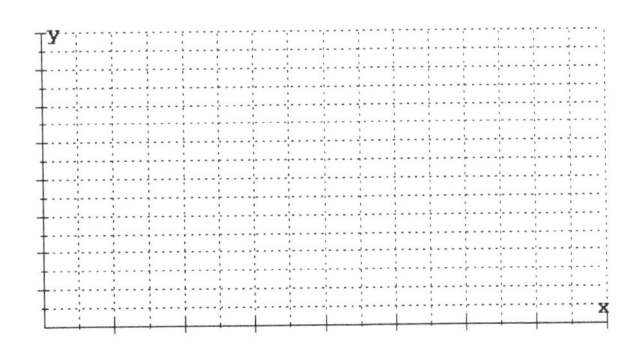

Copyright © 2013 Pearson Education, Inc.

173

29. Using the graph, estimate the number of miles driven if the cost of a one-day rental is $49. Verify your answer from the equation.

29._____

Name:
Instructor:

Date:
Section:

Chapter 4 GRAPHING LINEAR EQUATIONS AND INEQUALITIES

4.3 Graphing Linear Equations by Using Intercepts

Learning Objectives
A Find the *x*- and *y*-intercepts from the graph of a linear equation with two variables.
B Graph a linear equation with two variables by finding the *x*- and *y*-intercepts.
C Graph vertical and horizontal lines.

Getting Ready

Add, subtract, multiply, or divide the following integers:

1. $-6+5$ 1._____

2. $14-(-7)$ 2._____

3. $-3(4)$ 3._____

4. $-12+8$ 4._____

Substitute $x=-5$ and $y=0$ into the given expressions and evaluate:

5. $3x-5y$ 5._____

6. $-12x+7y$ 6._____

Solve the following equations:

7. $5x+3=8$ 7._____

8. $0-5y=20$ 8._____

Copyright © 2013 Pearson Education, Inc. 175

Plot the following ordered pairs on the rectangular coordinate system:

9. $(0, 5)$, $(-4, 2)$

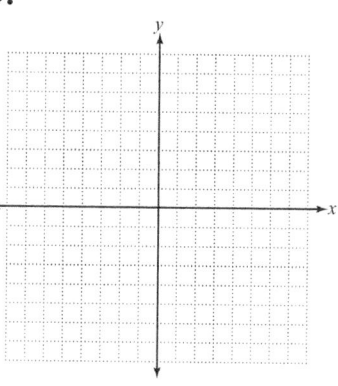

9.

10. Give the coordinates of the points on the graph.

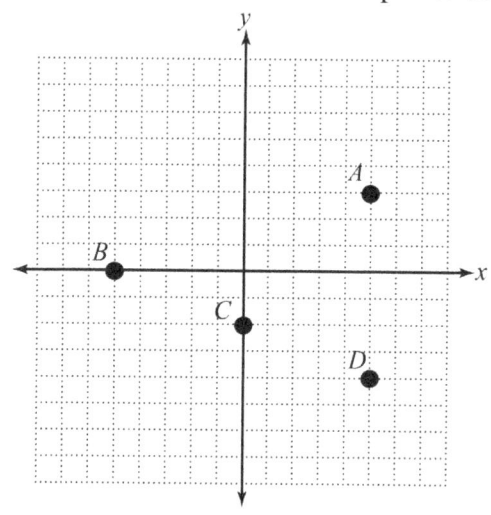

10. _____

Key Terms

Use the vocabulary terms listed below to complete each statement in exercises 1–6.

      x y *x*-intercept $x = k$ *y*-intercept $y = k$

1. The point where a graph intersects the *x*-axis is the _____.

2. The point where a graph intersects the *y*-axis is the _____.

3. To find the *x*-intercept let _____ = 0.

4. To find the *y*-intercept let _____ = 0.

5. A vertical line has an equation of the form _____.

6. A horizontal line has an equation of the form _____.

Name:
Instructor:

Date:
Section:

Practice Problems

Objective A Find the *x*- and *y*-intercepts from the graph of a linear equation with two variables.

Identify the x- and y-intercepts of the following.

1.

1._____

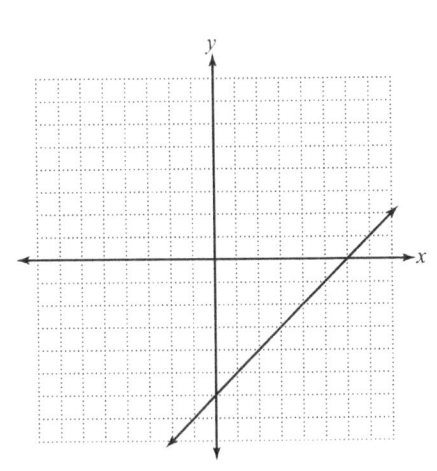

2.

2._____

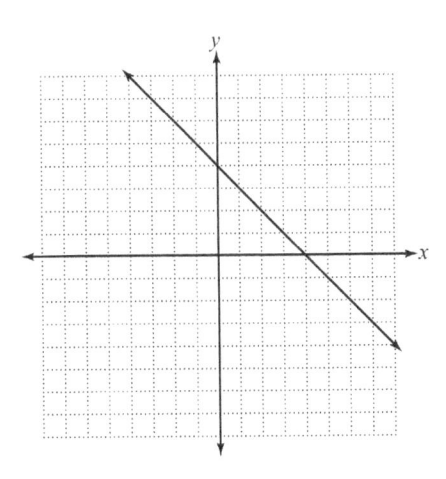

3.

3._____

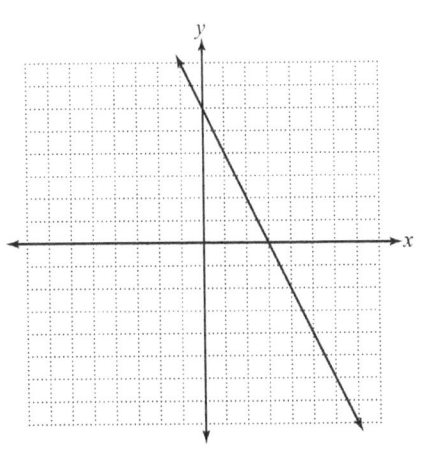

Copyright © 2013 Pearson Education, Inc.   177

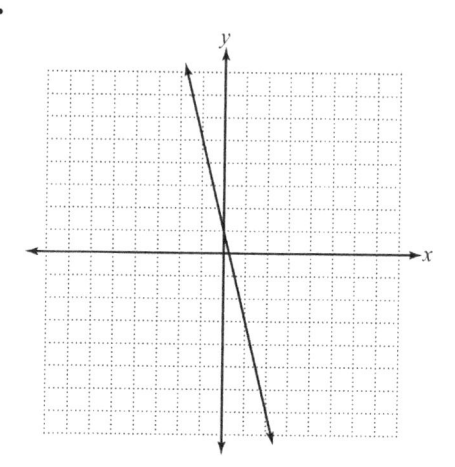

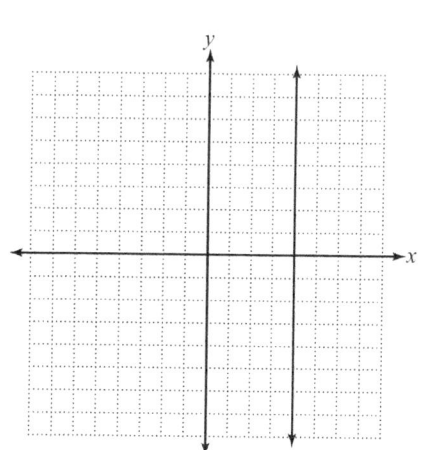

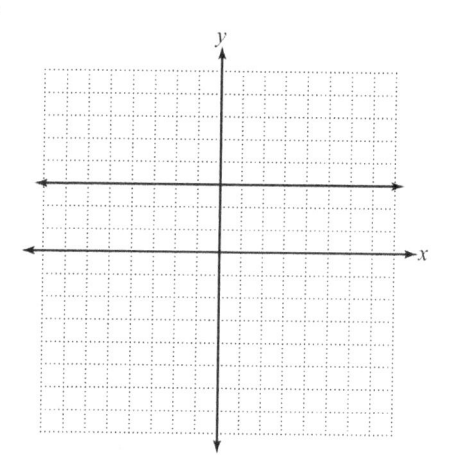

4.

5.

6.

4._____

5._____

6._____

Name:
Instructor:

Date:
Section:

Objective B Graph a linear equation with two variables by finding the *x*- and *y*-intercepts.

Draw the graph of the lines whose intercepts are given as follows.

7. $(6, 0), (0, -3)$

7.

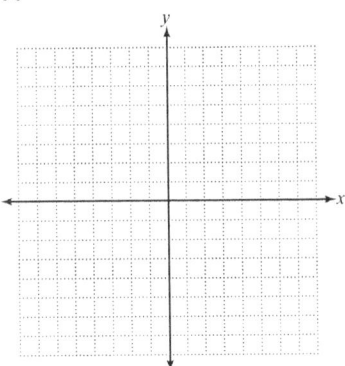

8. $(3, 0), (0, -5)$

8.

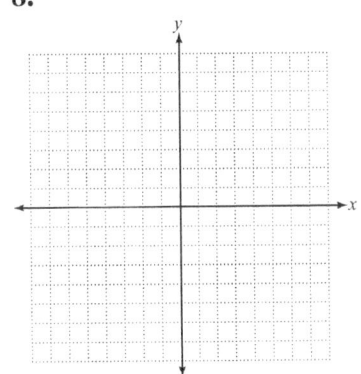

9. $(7, 0), (0, 6)$

9.

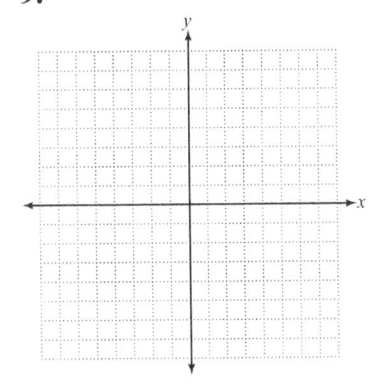

Copyright © 2013 Pearson Education, Inc. 179

10. $(-2, 0), (0, -2)$

Graph each of the following, using the intercept method.

11. $x + y = 7$

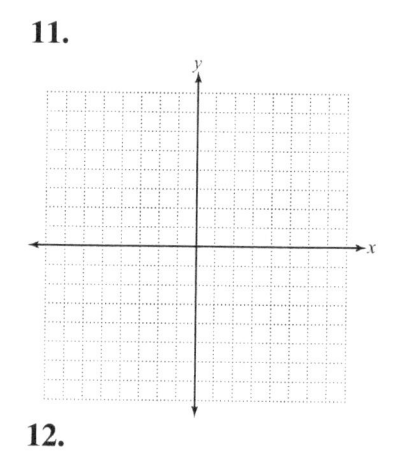

12. $3x - 4y = 12$

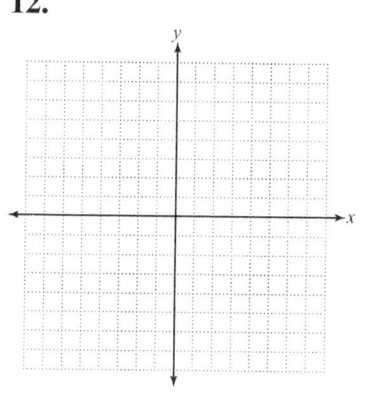

Name:
Instructor:

Date:
Section:

13. $4x - y = 8$

13.

14. $x - 6y = 3$

14.

15. $2x + 3y = 10$

15.

16. $4x + 7y = 8$

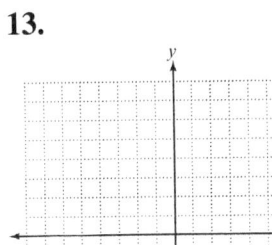

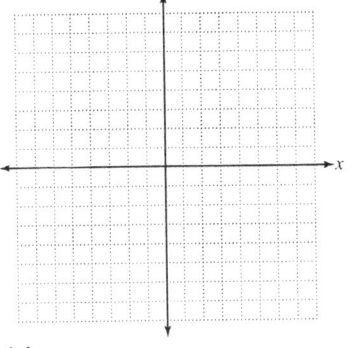

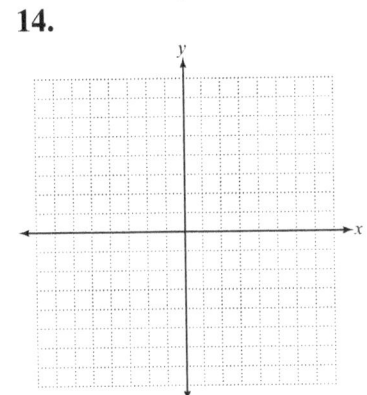

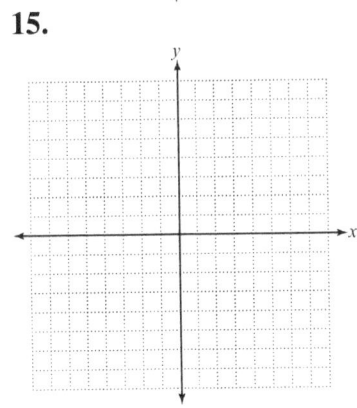

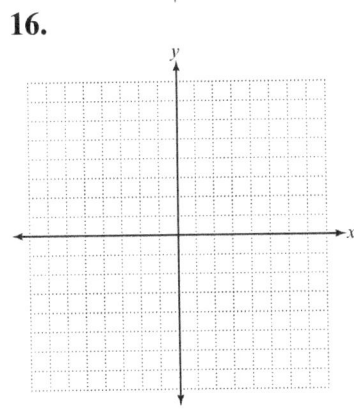

16.

Objective C Graph vertical and horizontal lines.

Graph each of the following.

17. $x = -2$

17.

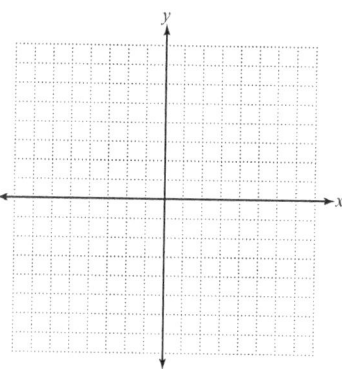

18. $y = 4$

18.

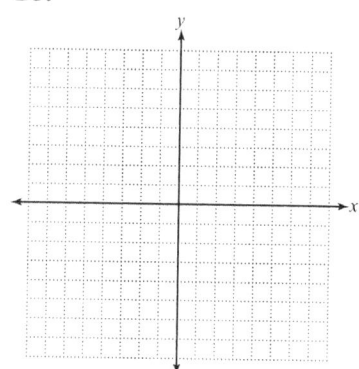

19. $y = -6$

19.

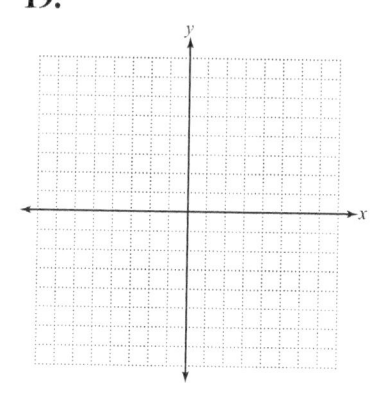

Name:
Instructor:

Date:
Section:

20. $x + 6 = 0$

20.

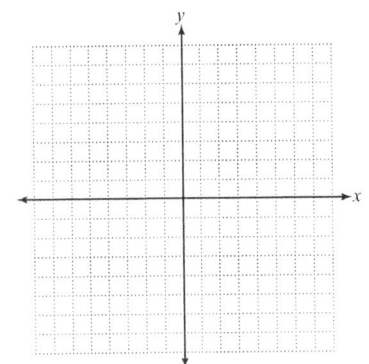

21. $y + 5 = 0$

21.

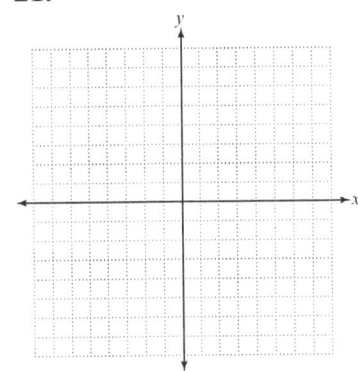

22. $x - 4 = 0$

22.

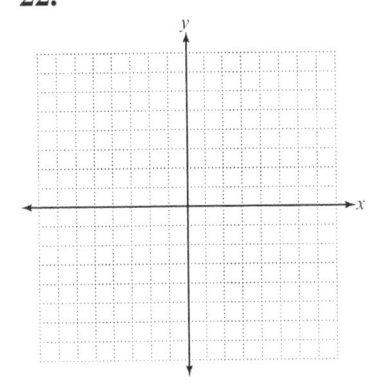

Name:
Instructor:
Date:
Section:

Chapter 4 GRAPHING LINEAR EQUATIONS AND INEQUALITIES

4.4 Slope of a Line

Learning Objectives
A Find the slope of a line by using the slope formula.
B Find the slope of a line from its graph.
C Graph a line when given a point on the line and the slope of the line.

Getting Ready

Add, subtract, multiply, or divide the following integers:

1. $-6-6$

2. $8-(-3)$

3. $-18(-3)$

4. $-24 \div (-8)$

1._____

2._____

3._____

4._____

Evaluate the following for the given values of the variables:

5. $\dfrac{x-y}{z-w}$: $x=10$, $y=4$, $z=2$, $w=1$

5._____

6. $\dfrac{x-y}{z-w}$: $x=-4$, $y=5$, $z=-8$, $w=-5$

6._____

Plot the following ordered pairs on the rectangular coordinate system:

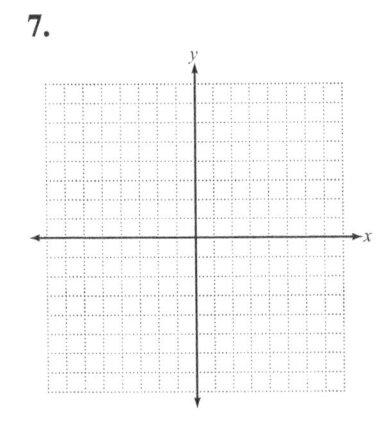

7. $(-6,0)$, $(-3,-3)$

7.

Key Terms

Use the vocabulary terms listed below to complete each statement in exercises 1–5.

| zero | undefined | slope | rises | falls |

1. The graph of a line with negative slope _____ from left to right.

2. The _____ of a line is the ratio of the change in y to the change in x as you go from one point on the line to any other point on the line.

3. The graph of a line with positive slope _____ from left to right.

4. The slope of a vertical line is _____.

5. The slope of a horizontal line is _____.

Practice Problems

Objective A Find the slope of a line by using the slope formula.

Find the slope of the line that contains each of the following pairs of points.

1. $(2,4), (-6,3)$ 1. _____

2. $(1,7), (0,6)$ 2. _____

3. $(-8,9), (10,15)$ 3. _____

4. $(12,7), (-8,7)$ 4. _____

5. $(0,16), (-14,0)$ 5. _____

6. $(3,4), (3,-9)$ 6. _____

Name:
Instructor:

Date:
Section:

7. $(9,-4), (6,8)$

7._____

8. $(-1,-4), (5,-9)$

8._____

Objective B   Find the slope of a line from its graph.

Find the slope of each of the following lines from the graph.

9.

9._____

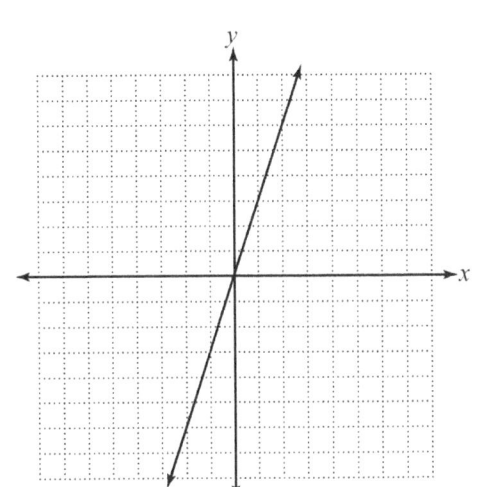

10.

10._____

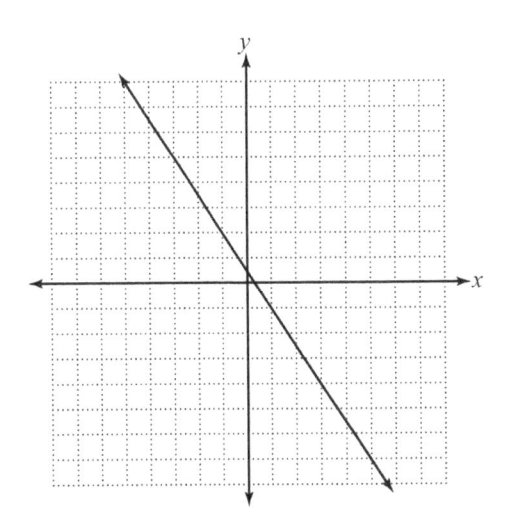

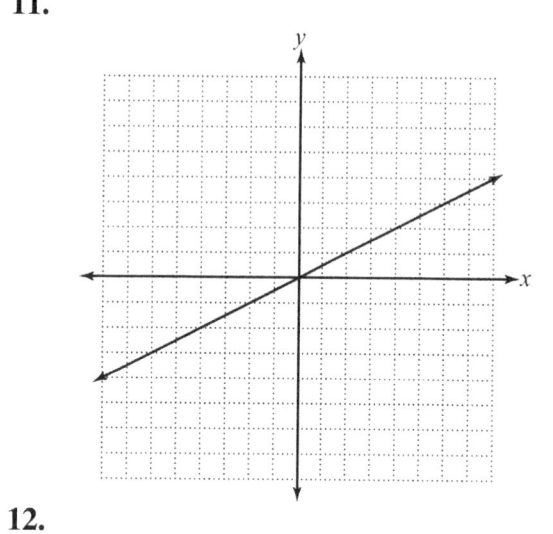

11.

11. _____

12.

12. _____

13.

13. _____

Name:
Instructor:

Date:
Section:

Objective C Graph a line when given a point on the line and the slope of the line.

Draw the graph of each of the following lines containing the given point and with the given slope.

14. $(-2, 4)$, $m = \dfrac{4}{5}$

14.

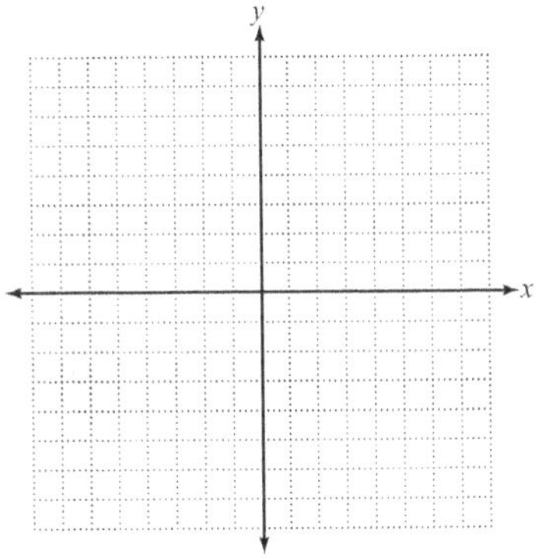

15. $(-2, -4)$, $m = \dfrac{2}{3}$

15.

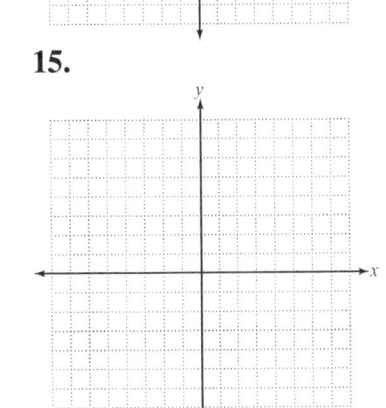

16. $(0, 5)$, $m = 2$

16.

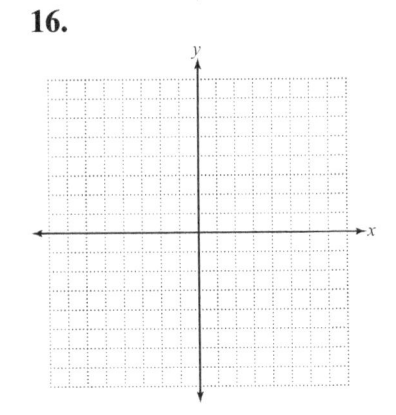

17. $(-4, 0)$, $m = -\dfrac{1}{3}$

17.

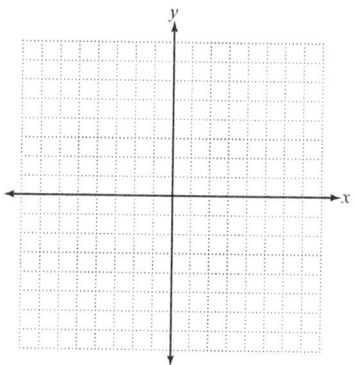

18. $(3, 6)$, undefined slope

18.

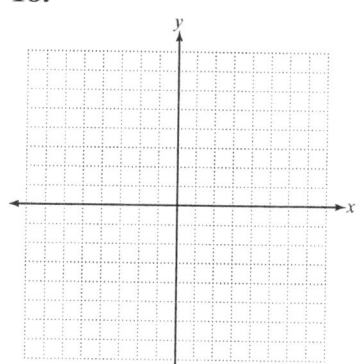

19. $(1, -4)$, $m = \dfrac{5}{2}$

19.

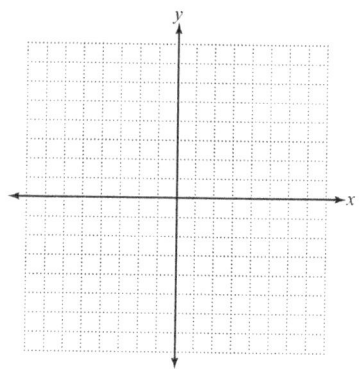

20. $(-2, 3)$, $m = 0$

20.

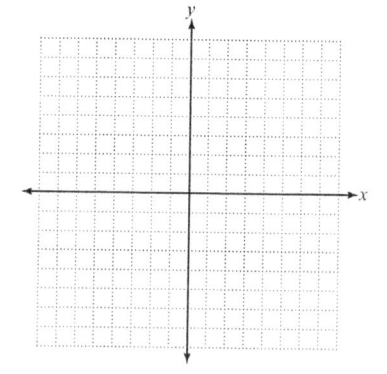

Name:
Instructor:

Date:
Section:

Find the slope of the graph of each of the following.

21. $x = -7$

21._____

22. $y = 8$

22._____

23. $y = 0$

23._____

24. $x = 0$

24._____

Name:　　　　　　　　　　　　　　　Date:
Instructor:　　　　　　　　　　　　　Section:

Chapter 4 GRAPHING LINEAR EQUATIONS AND INEQUALITIES

4.5 Slope-Intercept Form of a Line

Learning Objectives
A   Write an equation in slope-intercept form and find the slope and *y*-intercept from the equation.
B   Graph a line using the slope-intercept form.
C   Write the equation of a line, given the slope and *y*-intercept.
D   Find the solution of a linear equation from a graph.
E   Determine whether two lines are parallel or perpendicular.

Getting Ready

Add, subtract, multiply, or divide the following integers:

1. $-10-3$ 1._____

2. $7-(-5)$ 2._____

3. $-4(-13)$ 3._____

4. $-21 \div 7$ 4._____

5. Evaluate $\dfrac{x-y}{z-w}$: $x=-5$, $y=-8$, $z=-4$, $w=5$.   5._____

Plot the following ordered pairs on the rectangular coordinate system:

6. $(-5,-2)$, $(-4,2)$, $(1,0)$ 6.

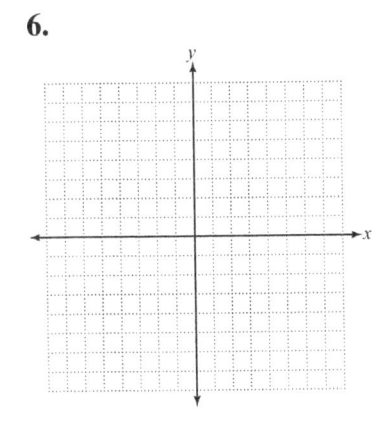

Copyright © 2013 Pearson Education, Inc.

Solve each of the following equations for y:

7. $5x + y = 7$

7._____

8. $6x - 7y = 14$

8._____

Key Terms

Use the vocabulary terms listed below to complete each statement in exercises 1–5.

    slope-intercept   **m**   **parallel**   **b**   **perpendicular**

1. Two nonvertical lines are _____ if and only if the product of their slopes is –1.

2. Two nonvertical lines are _____ if and only if they have equal slopes.

3. In the form $y = mx + b$, _____ is the y-intercept.

4. In the form $y = mx + b$, _____ is the slope.

5. $y = mx + b$ is called the _____ form of a line.

Practice Problems

Objective A, B   Write an equation in slope-intercept form and find the slope and y-intercept from the equation. Graph a line using the slope-intercept form.

Find the slope and the y-intercept of the line represented by each of the equations. Graph each equation by using the slope-intercept method.

1. $y = 3x + 4$

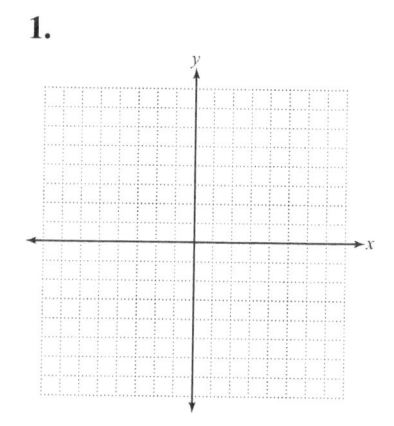

1.

194   Copyright © 2013 Pearson Education, Inc.

Name:
Instructor:

Date:
Section:

2. $y = -4x + 5$

2.

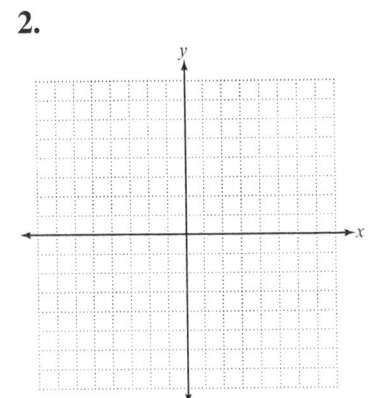

3. $y = -x + 5$

3.

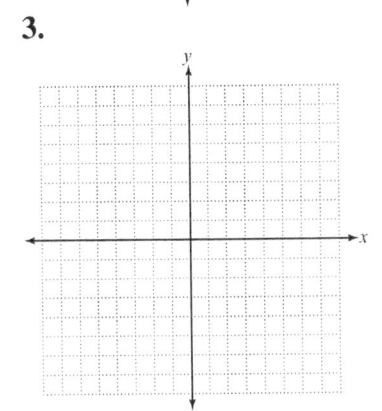

4. $y = \dfrac{1}{3}x + 4$

4.

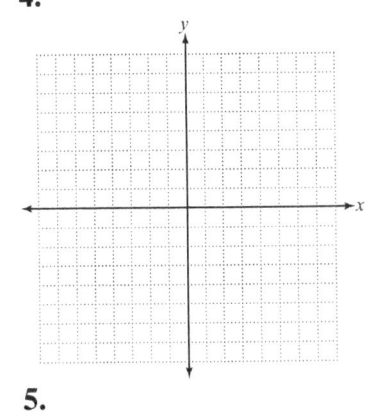

5. $y = -\dfrac{2}{7}x - 1$

5.

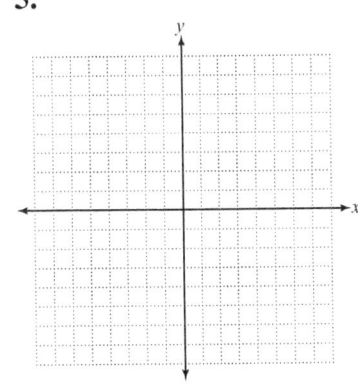

6. $y = -\dfrac{1}{4}x - 3$

6.

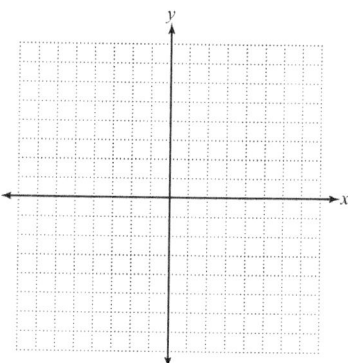

7. $2x + 5y = 15$

7.

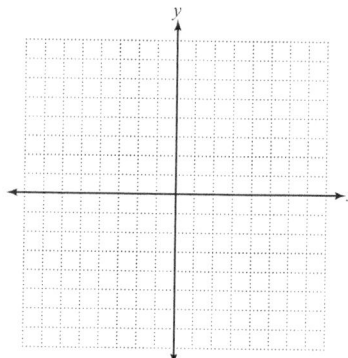

8. $-x + 6y = 3$

8.

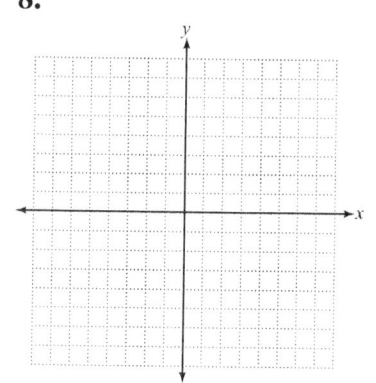

9. $3x - 4y = 9$

9.

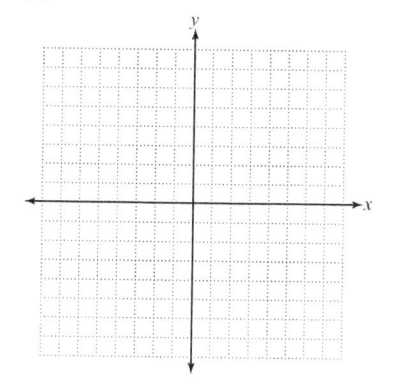

Name:
Instructor:

Date:
Section:

10. $6x + 7y = 12$

10.

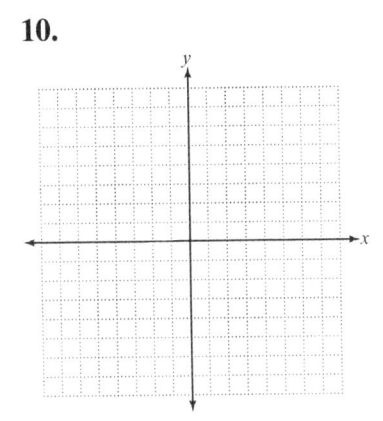

Objective C Write the equation of a line, given the slope and *y*-intercept.

Write the equations of the given lines. Express the answers in $y = mx + b$ form.

11. Slope is 3 and *y*-intercept is –5.

11._____

12. Slope is $-\dfrac{1}{6}$ and *y*-intercept is 9.

12._____

13. Slope is $-\dfrac{2}{3}$ and *y*-intercept is –8.

13._____

14. Slope is $\dfrac{3}{4}$ and *y*-intercept is $\dfrac{1}{8}$.

14._____

Objective D Find the solution of a linear equation from a graph.

Solve the given equation by using the given graph.

15. Solve $3x + 6 = 0$, using the given graph of $y = 3x + 6$.

15._____

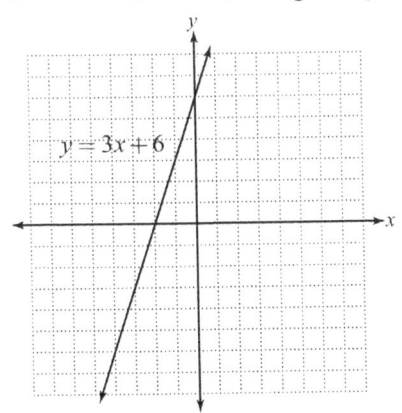

Copyright © 2013 Pearson Education, Inc.

197

16. Solve $\frac{2}{3}x - 4 = 0$, using the given graph of $y = \frac{2}{3}x - 4$.

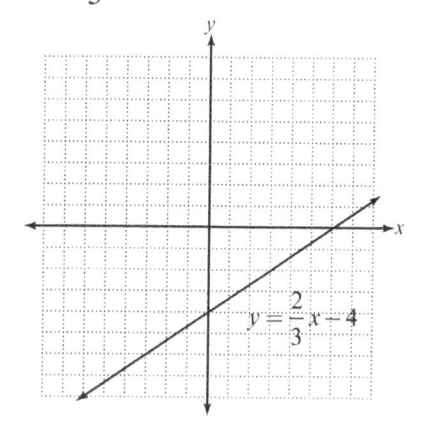

16._____

Objective E   Determine whether two lines are parallel or perpendicular.

Determine whether the lines containing the given pairs of points are parallel, perpendicular, or neither.

17. $L_1: (3,-5)$ and $(5,6)$;
    $L_2: (-9,4)$ and $(-7,15)$

17._____

18. $L_1: (2,-1)$ and $(3,7)$;
    $L_2: (2,5)$ and $(4,10)$

18._____

19. $L_1: (1,-6)$ and $(2,3)$;
    $L_2: (16,4)$ and $(7,5)$

19._____

Name:
Instructor:

Date:
Section:

20. $L_1: (-8,-7)$ and $(0,6)$;

$L_2: (5,3)$ and $(13,-10)$

20._____

Determine whether the graphs of the lines with the given equations are parallel, perpendicular, or neither.

21. $L_1: 3x - y = 8$;

$L_2: 3x + y = -7$

21._____

22. $L_1: 4x - 6y = 14$;

$L_2: 6x - 4y = -2$

22._____

23. $L_1: x = 3$;

$L_2: y = -8$

23._____

24. $L_1: x + 5 = 0$;

$L_2: x = 11$

24._____

25. $L_1: 6x+12y=19;$
$L_2: 4x+8y=7$

25._____

Name:
Instructor:

Date:
Section:

Chapter 4 GRAPHING LINEAR EQUATIONS AND INEQUALITIES

4.6 Point-Slope Form of a Line

Learning Objectives
A   Write the equation of a line that contains a given point and has a given slope.
B   Write the equation of a line that contains two given points.
C   Write the equations of vertical and horizontal lines.
D   Write the equation of a line that contains a given point and is parallel or perpendicular to a line whose equation is given.

Getting Ready

Add, subtract, multiply, or divide the following integers:

1. $13 - (-9)$ 1._____

2. $-56 \div 8$ 2._____

Evaluate the following for the given value(s) of the variable(s):

3. $\dfrac{3}{4}x + 9$: $x = 8$ 3._____

4. $\dfrac{x-y}{z-w}$: $x = 8$, $y = 4$, $z = -2$, $w = -1$ 4._____

5. Simplify $7(x-2)$, using the distributive property. 5._____

6. Write $4(y-3) = 2 \cdot \dfrac{2}{3}(x-2)$ in $ax + by = c$ form. 6._____

7. Solve $4(y-3) = 2 \cdot \dfrac{2}{3}(x-2)$ for y. 7._____

Copyright © 2013 Pearson Education, Inc.

8. Find the slope of the line that contains $(-8,2)$, $(-4,6)$.   8._____

9. Find the slope of the graph of $4x-5y=10$.   9._____

State whether the following pairs of lines are parallel, perpendicular, or neither:

10. $3x+y=-2$, $-2x-y=4$ 10._____

11. $-5x-2y=8$, $2x-5y=8$ 11._____

12. Find the value of y: $y=\dfrac{4}{5}(2000)+600$.   12._____

Key Terms

Use the vocabulary terms listed below to complete each statement in exercises 1–5.

point-slope formula   **equal**   **negative reciprocals**
vertical   **horizontal**

1. Parallel lines have_____ slopes.

2. The graph of a linear equation with two variables of the form $x=k$ is a_____ line.

3. The graph of a linear equation with two variables of the form $y=k$ is a_____ line.

4. The slopes of perpendicular lines are_____ of each other.

5. $y-y_1=m(x-x_1)$ is called the _____ of a line.

Name:
Instructor:

Date:
Section:

Practice Problems

Objective A Write the equation of a line that contains a given point and has a given slope.

Write the equation of each of the following lines that contain the given point and have the given slope in $ax+by=c$ form with a and b integers.

1. $(7,5)$, $m=6$

    1._____

2. $(-4,9)$, $m=-2$

    2._____

3. $(-1,-10)$, $m=\dfrac{1}{5}$

    3._____

4. $(9,0)$, $m=-\dfrac{2}{7}$

    4._____

5. $(-6,9)$, $m=\dfrac{3}{5}$

    5._____

6. $(8,7)$, horizontal

    6._____

7. $(-9, 11)$, vertical   7._____

8. $(-1, -2)$, $m = 0$ 8._____

9. $(2, -4)$, undefined slope   9._____

Objectives B, C Write the equation of a line that contains two given points. Write the equations of vertical and horizontal lines.

Write the equation of the line that contains the given pairs of points. Express answers in slope-intercept form.

10. $(3, 8)$, $(7, -6)$ 10._____

11. $(-1, -2)$, $(-4, 3)$ 11._____

12. $(5, 9)$, $(-16, 9)$ 12._____

Name:
Instructor:

Date:
Section:

13. $(4,0), (-5,-2)$ 13._____

14. $(4,12), (4,-8)$ 14._____

Objective D Write the equation of a line that contains a given point and is parallel or perpendicular to a line whose equation is given.

Write the equations of the following lines in $ax+by=c$ form with a, b, and c integers.

15. Containing $(-2,-5)$ and parallel to the graph of $7x-3y=12$ 15._____

16. Containing $(-7,0)$ and parallel to the graph of $5x+8y=9$ 16._____

17. Containing $(5,1)$ and perpendicular to the graph of $8x-3y=4$ 17._____

18. Containing $(3,1)$ and perpendicular to the graph of $x-3y=8$ 18._____

Copyright © 2013 Pearson Education, Inc.

19. containing $(11,14)$ and parallel to the graph of $y=7$

19._____

20. containing $(-6,-2)$ and perpendicular to the graph of $y=3$

20._____

21. containing $(0,-1)$ and parallel to the graph of $x=15$

21._____

22. containing $(-6,7)$ and perpendicular to the graph of $x=4$

22._____

23. The annual tuition at a dance studio depends on the number of 1-hour classes taken. The ordered pair $(3,720)$ means that 3 1-hour classes cost a total of $720 per year, and $(5,1200)$ means that 5 1-hour classes cost a total of $1200.
a. Write a linear equation that gives the cost, y, in terms of the number of 1-hour classes, x.
b. Use the equation in part a to find the cost for taking 8 1-hour classes.

23a._____

b._____

206   Copyright © 2013 Pearson Education, Inc.

Name: Date:
Instructor: Section:

Chapter 4 GRAPHING LINEAR EQUATIONS AND INEQUALITIES

4.7 Graphing Linear Inequalities with Two Variables

Learning Objectives
A   Graph linear inequalities with two unknowns.

Getting Ready

Add, subtract, multiply, or divide the following integers:

1. $-8-4$ 1._____

2. $5-(-7)$ 2._____

3. $-6(-11)$ 3._____

4. $-36 \div (-6)$ 4._____

Substitute $x=-5$ and $y=6$ into the given expressions and evaluate.

5. $-3x+4y$ 5._____

6. $2x-5y$ 6._____

Graph each of the following linear equations:

7. $6x + 4y = 20$

7.

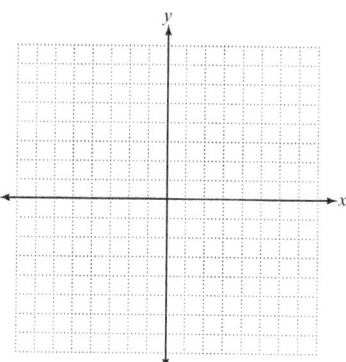

8. $3x - y = 0$

8.

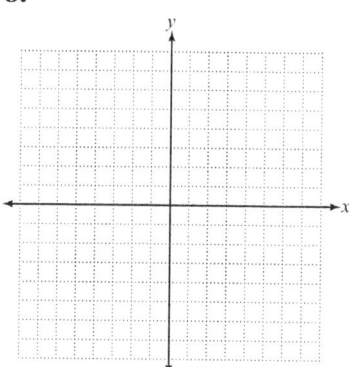

9. $y = -5$

9.

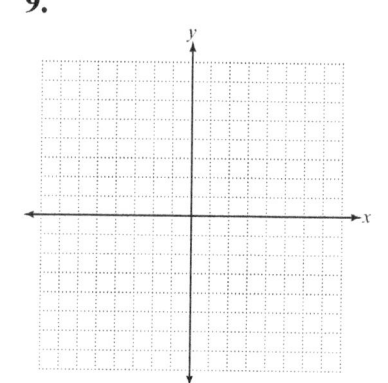

10. $x = 4$

10.

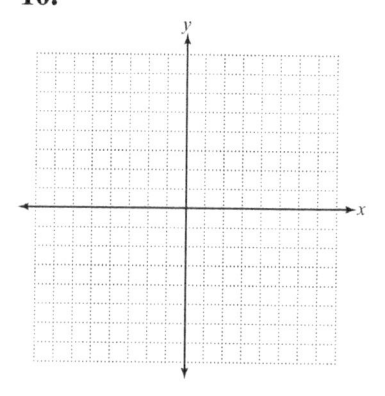

Name:
Instructor:

Date:
Section:

Key Terms

Use the vocabulary terms listed below to complete each statement in exercises 1–3.

linear inequality **dashed** **solid**

1. $ax + by < c$ is called a _____.

2. When graphing the solution of $ax + by \leq c$ draw the line $ax + by = c$ as a _____ line.

3. When graphing the solution of $ax + by < c$ draw the line $ax + by = c$ as a _____ line.

Practice Problems

Objective A   Graph linear inequalities with two unknowns.

Determine whether the given point is a solution of the given inequality.

1. $y < 5x - 4$; $(1, 2)$

    1._____

2. $2x + y \geq 7$; $(2, 3)$

    2._____

3. $6x - 8y > 9$; $(-1, 6)$

    3._____

Graph the given linear inequalities with two variables.

4. $y \leq x + 3$

    4.

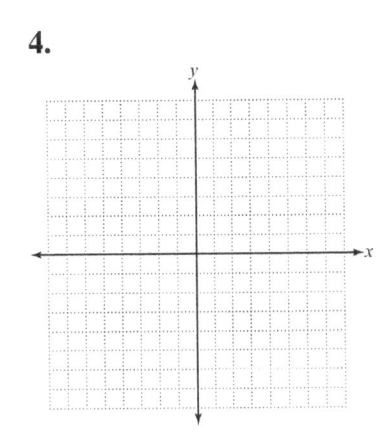

5. $y \leq x - 5$

5.

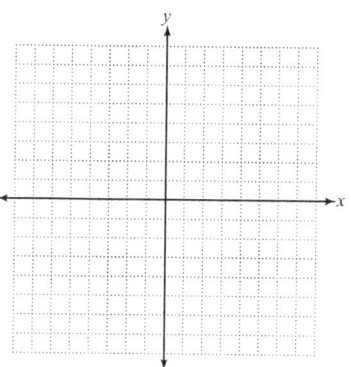

6. $y \geq 2x + 4$

6.

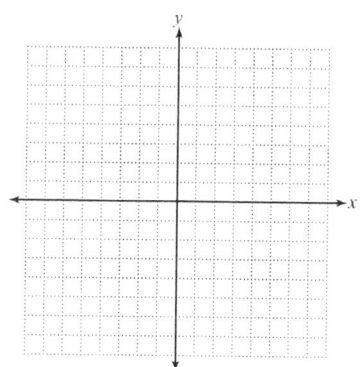

7. $y \geq \dfrac{1}{4}x + 2$

7.

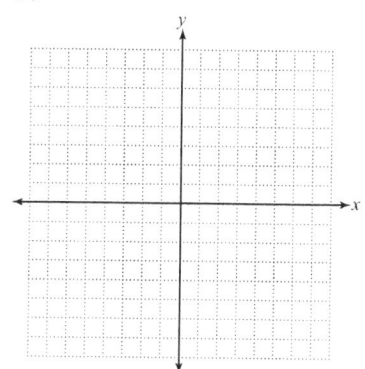

8. $y > \dfrac{3}{2}x - 1$

8.

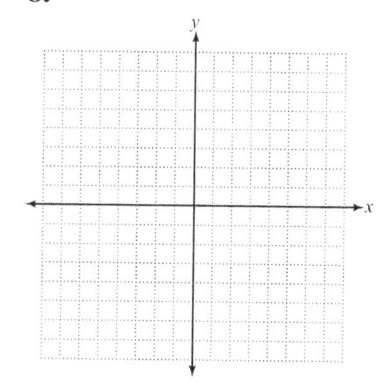

Name:
Instructor:

Date:
Section:

9. $x + 5y > 5$

9.

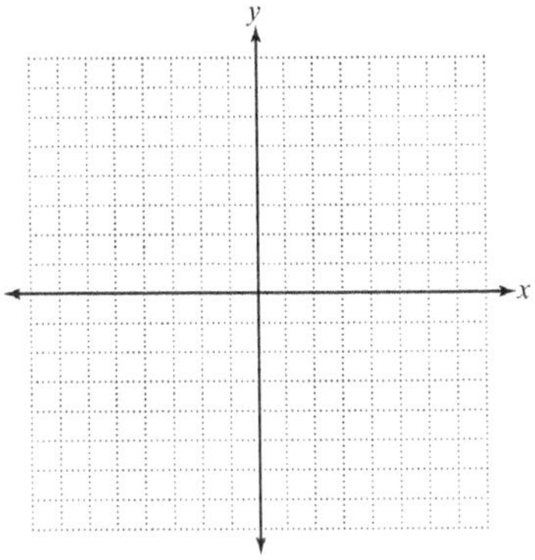

10. $3x - 4y \geq 8$

10.

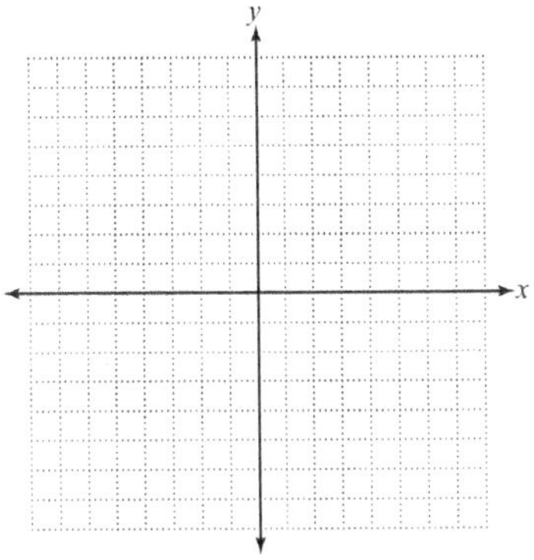

11. $y < 4x$

11.

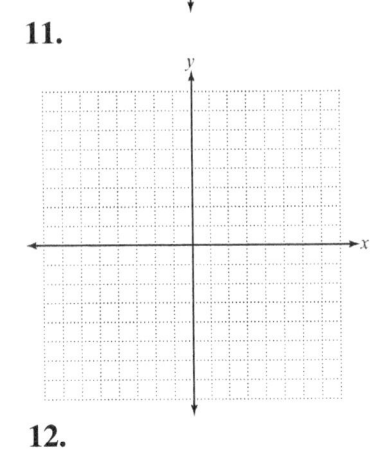

12. $y \geq 3x$

12.

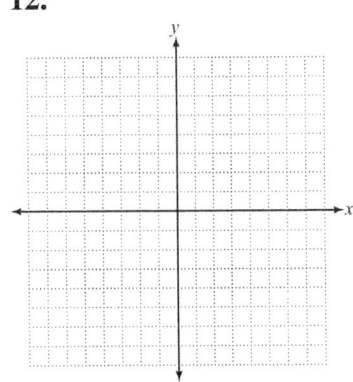

13. $x \geq 3$

13.

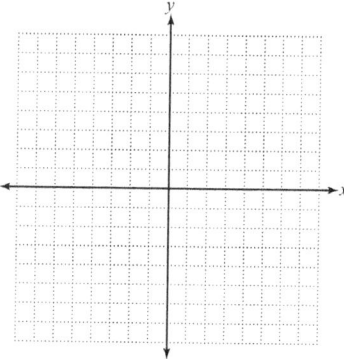

14. $y < 5$

14.

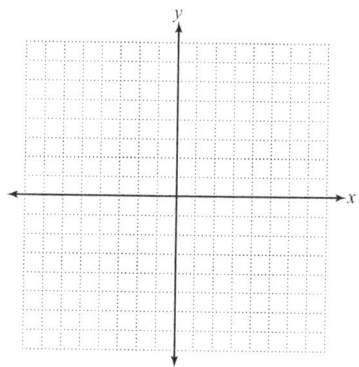

15. $x < -2$

15.

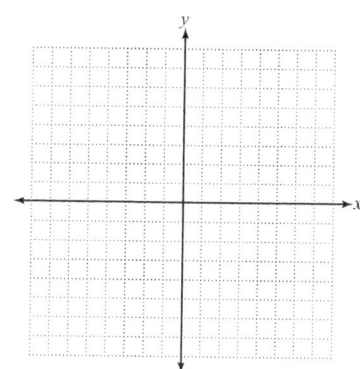

16. $y + 6 \leq 0$

16.

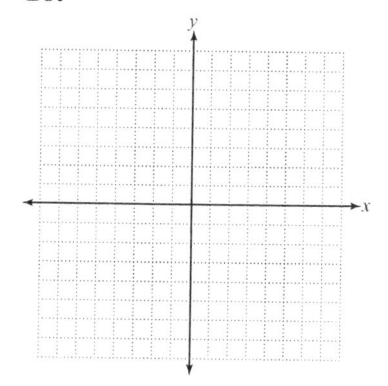

Name: Date:
Instructor: Section:

17. $y - 5 > 0$

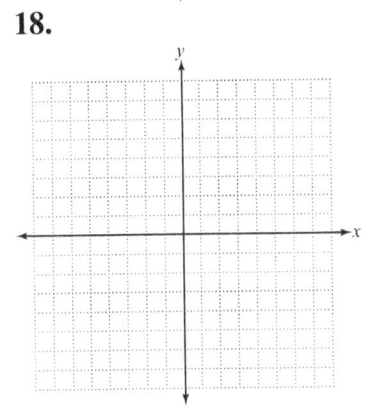

17.

18. $x + 3 \geq 0$

18.

Name:
Instructor:

Date:
Section:

Chapter 4 GRAPHING LINEAR EQUATIONS AND INEQUALITIES

4.8 Relations and Functions

Learning Objectives
A   Determine the domain and range of a relation.
B   Recognize functions when they are given as a set of ordered pairs, as a graph, or as an equation.
C   Use functional notation.

Getting Ready

Plot the following ordered pairs on the rectangular coordinate system:

1. $(-5,-2)$, $(4,-6)$, $(3,0)$, $(-1,-7)$

1.

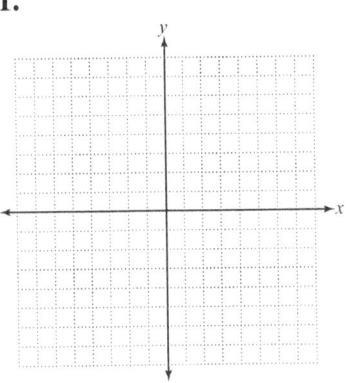

Decide whether the graphed line is horizontal, vertical, or neither.

2.

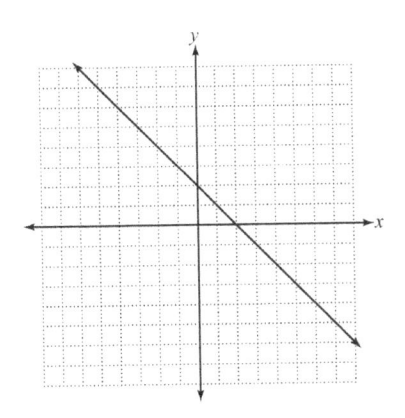

2._____

Copyright © 2013 Pearson Education, Inc.   215

3.

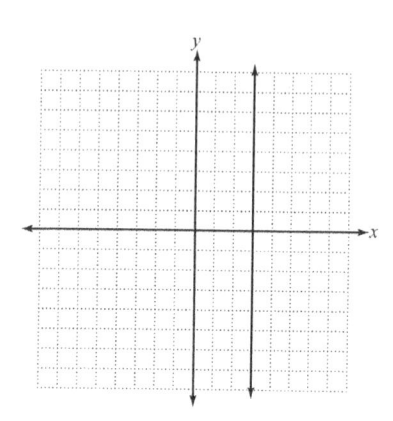

3._____

Evaluate the following expressions using the indicated value of the variable:

4. $5x-8$, $x=3$

4._____

5. x^2-3, $x=5$

5._____

Find the following sum or product as indicated:

6. $-6+3$

6._____

7. $9(-6)$

7._____

Key Terms

Use the vocabulary terms listed below to complete each statement in exercises 1–7.

domain	range	elements	relation
function	evaluating	vertical line	

1. The objects that make up a set are called the _____ of the set.

2. A(n) _____ is any set of ordered pairs.

3. The set of all first components in a set of ordered pairs is called the _____.

216   Copyright © 2013 Pearson Education, Inc.

Name: Date:
Instructor: Section:

4. The set of all second components in a set of ordered pairs is called the
   _____.

5. A(n) _____ is a relation in which every first component is paired with exactly one second component.

6. If any _____ intersects the graph of a relation in more than one point, then that relation is not a function.

7. When finding the value of $f(x)$ for a specific value of x, we are _____ the function.

Practice Problems

Objective A Determine the domain and range of a relation.

Determine which of the following relations are functions and give the domain and range of each.

1. $\{(-3,4), (0,6), (1,5)\}$ 1._____

2. $\{(0,1), (1,0), (-1,0), (2,-9)\}$ 2._____

3. $\{(3,6), (-3,5), (5,2), (3,8)\}$ 3._____

Objective B Recognize functions when they are given as a set of ordered pairs, as a graph, or as an equation.

Determine whether each of the following is the graph of a function.

4.   4._____

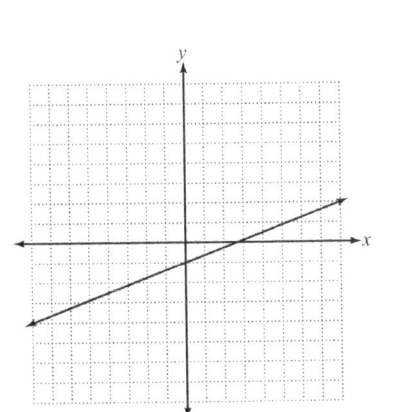

Copyright © 2013 Pearson Education, Inc.   217

5.

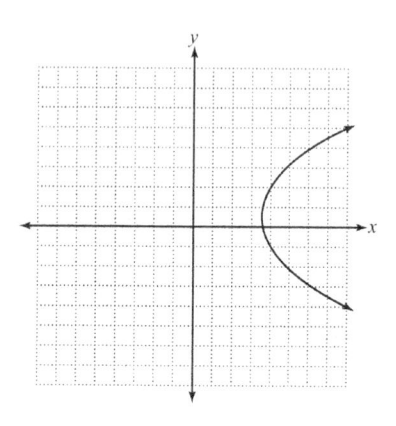

[graph of sideways parabola opening right]

5._____

6.

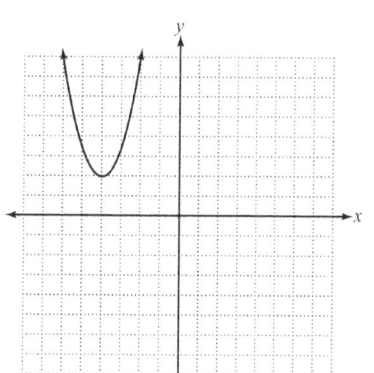

[graph of upward parabola in second quadrant]

6._____

7.

[zigzag graph]

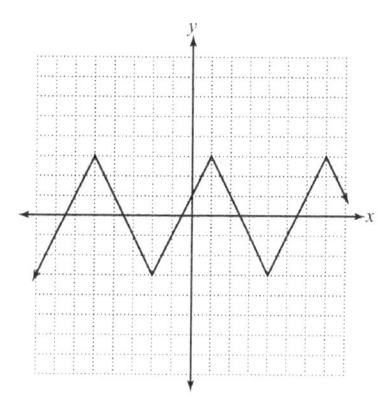

7._____

Determine whether each of the following define y as a function of x.

8. $y = -7x + 5$

8._____

9. $y^2 = 2x + 3$

9._____

Name:
Instructor:

Date:
Section:

10. $x + y = 9$

10._____

11. $y = x^2 - 5$

11._____

12. $x^2 + y^2 = 9$

12._____

13. $y = |4x - 6|$

13._____

14. $y \geq 7x + 8$

14._____

Objective C Use functional notation.

Given $f(x) = -4x + 6$, *find the following and represent each result as an ordered pair.*

15. $f(-1)$

15._____

16. $f(0)$

16._____

17. $f(3)$

17._____

18. $f(t)$

18._____

Given $g(x) = x^2 + 3$, find the following and represent each result as an ordered pair.

19. $g(4)$ 19._____

20. $g(-6)$ 20._____

21. $g(0)$ 21._____

22. $g(b)$ 22._____

The cost, $C(x)$, as a function of the number of items produced, x, is given by $C(x) = 50x + 600$.

23. Find the cost of producing 20 items.   23._____

24. What does $C(50) = 3100$ mean?   24._____

Name: Date:
Instructor: Section:

Chapter 5   FACTORS, DIVISORS, AND FACTORING

5.1   Prime Factorization and Greatest Common Factor

Learning Objectives
A   Recognize prime and composite numbers.
B   Determine whether a number is divisible by 2, 3, or 5.
C   Find the prime factorization of composite numbers.
D   Find the greatest common factor for two or more natural numbers.
E   Find the greatest common factor for two or more monomials that contain variables.

Getting Ready

Find the product or quotient.

1. $108 \div 4$ 1._____

2. $378 \div 6$ 2._____

Write the following in exponential form:

3. $2 \cdot 2 \cdot 3 \cdot 3 \cdot 4 \cdot 4 \cdot 4$ 3._____

4. $3 \cdot 5 \cdot 5 \cdot 5 \cdot 7 \cdot 7$ 4._____

Evaluate the given expressions.

5. $3 \cdot 4^2$ 5._____

6. $2^2 \cdot 3 \cdot 5$ 6._____

Key Terms
Use the vocabulary terms listed below to complete each statement in exercises 1–6.

   prime   composite   factorization   greatest common factor
   prime factorization   divisible

1. A _____ number is a natural number with more than two factors.

Copyright © 2013 Pearson Education, Inc.   221

2. The number *a* is _____ by the number *b* if the remainder is 0 when *a* is divided by *b*.

3. A _____ number is a natural number with two, and only two, factors.

4. The _____ for a group of numbers is the largest number that will divide evenly into each number of the group.

5. A _____ of a number *c* consists of writing *c* as the product of two or more numbers.

6. There is one and only one _____ of any composite number.

Practice Problems

Objective A Recognize prime and composite numbers.

Classify each of the following as prime or composite.

1. 43   1._____

2. 93   2._____

3. 31   3._____

4. 187   4._____

List all natural number factors (divisors) of each of the following.

5. 68   5._____

6. 76   6._____

7. 22   7._____

Name: Date:
Instructor: Section:

Objective B Determine whether a number is divisible by 2, 3, or 5.

Determine whether each of the following is divisible by 2, 3, 5 or none of these.

8. 45   8._____

9. 122   9._____

10. 81   10._____

11. 93   11._____

12. 582   12._____

13. 41   13._____

Objective C Find the prime factorization of composite numbers.

Find the prime factorization of each of the following.

14. 52   14._____

15. 1575   15._____

16. 132   16._____

17. 1309   17._____

18. 540   18._____

Objective D   Find the greatest common factor for two or more natural numbers.

Find the greatest common factor (GCF) of each of the following.

19. 35 and 55

19._____

20. 33 and 52

20._____

21. 72 and 112

21._____

22. 4, 24, and 36

22._____

23. 14, 21, and 70

23._____

Objective E   Find the greatest common factor for two or more monomials that contain variables.

Find the greatest common factor (GCF) of each of the following.

24. $x^3 y^2$ and $x^6 y^{10}$

24._____

25. ab^5 and $a^2 b^3$

25._____

26. $m^4 n^7$ and $m^3 n^6$

26._____

27. $12w^2 z^5$ and $18w^4 z^3$

27._____

28. $6ad^2$ and $39a^2 d$

28._____

29. $120 p^5 q^{12}$, $42 pq^8$, and $28 p^3 q$

29._____

30. $22x^3 yz^5$, $154 x^7 y^3 z^4$, and $33 x^2 y^6 z^9$

30._____

Name: Date:
Instructor: Section:

Chapter 5 FACTORS, DIVISORS, AND FACTORING

5.2 Factoring Polynomials with Common Factors and by Grouping

Learning Objectives
A Factor a polynomial by removing the greatest common factor.
B Factor by grouping.

Getting Ready

Find the following products:

1. $3(3a+4b)$ 1._____

2. $(-4)(6x+5y)$ 2._____

3. $2y^2(y^3-3z)$ 3._____

4. $b^3y^2(b+y)$ 4._____

5. $ab(3a^2-5ab-6b^2)$ 5._____

Find the following quotients:

6. $\dfrac{9x^5y^4}{3x^2y^4}$ 6._____

7. $\dfrac{12a^3b^4}{6a^3b}$ 7._____

Find the following products:

8. $(-4)(6)$ 8._____

9. $(-5)(-8)$ 9._____

Copyright © 2013 Pearson Education, Inc. 225

Key Terms

Use the vocabulary terms listed below to complete each statement in exercises 1–4.

factorization   **factoring by grouping**
distributive property   **greatest common factor**

1. $ab + ac = a(b+c)$ demonstrates the _____.

2. _____ can be checked by showing that the product of the factors is the same as the original polynomial.

3. One method for factoring a polynomial of more than 3 terms is _____.

4. One of the first steps in factoring is to remove the _____.

Practice Problems

Objective A   Factor a polynomial by removing the greatest common factor.

Factor the following by removing the greatest common factor.

1. $mw - nw$ 1._____

2. $5x - 75$ 2._____

3. $x^5 + x^3$ 3._____

4. $12y^5 + 6y^2$ 4._____

5. $32a^4 - 16a^3$ 5._____

6. $15x^3 + 45x^2$ 6._____

7. $48z^3 - 12z^6$ 7._____

8. $a^2b^5 - ab^6$ 8._____

Name:
Instructor:

Date:
Section:

9. $pq^4 + p^2q$

9._____

10. $3x^2y^3 - 6x^4y^7$

10._____

11. $2x^2 - 6x + 18$

11._____

12. $4a^3 + 16a^2 - 64$

12._____

13. $6m^2 + 4m + 8$

13._____

14. $9w^4 + 3w^2 - 24$

14._____

15. $7x^2y^3 + 21xy^5 - 35x^3y^4$

15._____

Factor the following by removing a negative common factor.

16. $-12c + 15d$

16._____

17. $-44z + 33$

17._____

18. $-3x^3 + 6x^2 - 12$

18._____

Factor the following by removing the common binomial factor.

19. $a(c+d) - 2(c+d)$

19._____

20. $m(p-q) - n(p-q)$

20._____

21. $x(y+8)+z(y+8)$

21._____

Objective B   Factor by grouping.

Factor each of the following by grouping.

22. $ac-ad+bc-bd$

22._____

23. $xy+2y+4x+8$

23._____

24. $a^2-6a-a+6$

24._____

25. $21w^2-7wz+3wz-z^2$

25._____

26. $2mn-12m+3n-18$

26._____

27. $2ab+6ad+bc+3cd$

27._____

28. $14xw-21xz-2wy+3yz$

28._____

29. $15x^2-3xy+10x-2y$

29._____

30. $12x^2-9xy+8xy-6y^2$

30._____

Name: Date:
Instructor: Section:

Chapter 5 FACTORS, DIVISORS, AND FACTORING

5.3 Factoring General Trinomials with Leading Coefficients of One

Learning Objectives
A Factor a general trinomial whose first coefficient is one when written in descending order.

Getting Ready

Find the following products, using FOIL:

1. $(y+4)(y-3)$ 1._____

2. $(x-7)(x-6)$ 2._____

3. Find the product: $xy \cdot x^3 \cdot y^2$ 3._____

Find the following sums:

4. $-4a+7a$ 4._____

5. $8ab-2ab$ 5._____

Find the following products:

6. $8(-6)$ 6._____

7. $(-9)(-4)$ 7._____

Factor the following:

8. $3x^2-3x-6$ 8._____

9. $5m^2n^2-40mn^2+75n^2$ 9._____

Copyright © 2013 Pearson Education, Inc.   229

Key Terms

Use the vocabulary terms listed below to complete each statement in exercises 1–4.

    prime   trinomial   positive   negative

1. If the last term of a factorable trinomial is _____ both binomial factors will have the same sign as the middle sign of the trinomial.

2. If the last term of a factorable trinomial is _____ the signs of the binomial factors will be opposites.

3. A polynomial that cannot be factored is called _____.

4. A polynomial with three terms is called a _____.

Practice Problems

Objective A Factor a general trinomial whose first coefficient is one when written in descending order.

Factor the following, if possible; if the polynomial is not factorable, write "prime."

1. $z^2 - 4z - 96$ 1._____

2. $y^2 + 4y + 3$ 2._____

3. $b^2 + 5b - 6$ 3._____

4. $p^2 + 13p + 40$ 4._____

5. $a^2 - 5a - 4$ 5._____

6. $d^2 - 7d + 12$ 6._____

Name:
Instructor:

Date:
Section:

7. $c^2 - 3c - 88$

7._____

8. $m^2 + 13m + 42$

8._____

9. $x^2 + 22x + 40$

9._____

10. $y^2 - 20y + 91$

10._____

11. $x^2 + 6x - 5$

11._____

12. $q^2 + 17q + 60$

12._____

13. $m^2 + 11mn + 18n^2$

13._____

14. $p^2 - 8pq + 15q^2$

14._____

15. $x^2 + xy - 2y^2$

15._____

16. $a^2 + 9ab + 14b^2$

16._____

17. $c^2 - 6cd - 7d^2$

17._____

18. $w^2 + 6wz + 5z^2$

18._____

19. $r^2 + rt - 20t^2$

19._____

20. $x^2 + 20xy + 19y^2$

20._____

21. $5x^2 + 5x - 10$

21._____

22. $2a^2 - 2a - 12$

22._____

23. $4a^2 + 44a + 120$

23._____

24. $11w^2 + 44w + 33$

24._____

25. $10y^2 - 130y + 420$

25._____

26. $8n^2 - 64n + 120$

26._____

27. $a^3 + 13a^2 + 36a$

27._____

28. $x^5 - 15x^4 + 50x^3$

28._____

29. $p^4 - 6p^3 - 7p^2$

29._____

30. $5m^3 + 65m^2 + 200m$

30._____

Name: Date:
Instructor: Section:

Chapter 5 FACTORS, DIVISORS, AND FACTORING

5.4 Factoring General Trinomials with Leading Coefficients Other Than One

Learning Objectives
A Factor trinomials whose leading coefficient is a natural number other than one.

Getting Ready

Factor the following:

1. $x^2 + 2x - 63$ 1._____

2. $x^2 + 5x + 4$ 2._____

3. $w^4 - 3w^3$ 3._____

4. $bc^5 + bc^3 - 5bc^2$ 4._____

Key Terms

Use the vocabulary terms listed below to complete each statement in exercises 1–3.

      ac   leading   greatest common factor

1. When a polynomial is written in descending order, the coefficient of the first term is the _____ coefficient.

2. In any factorization, the first thing we always look for is a(n) _____.

3. One method of factoring trinomials is trial-and-error, another method is the _____ method.

Practice Problems

Objective A Factor trinomials whose leading coefficient is a natural number other than one.

Factor each expression completely. If the polynomial will not factor, write "prime."

1. $2a^2 - a - 15$ 1._____

2. $4x^2 + 13x + 3$

2._____

3. $6m^2 + 3m + 5$

3._____

4. $7y^2 - 30y + 8$

4._____

5. $5b^2 + 19b + 12$

5._____

6. $4c^2 + 15c + 9$

6._____

7. $6p^2 + 17p + 5$

7._____

8. $4b^2 - 2b - 7$

8._____

9. $6a^2 - 23a + 20$

9._____

10. $4a^2 + 16ab + 15b^2$

10._____

11. $3m^2 + 20m - 7$

11._____

12. $9n^2 - 5n + 12$

12._____

Name:
Instructor:

Date:
Section:

13. $36m^2 + 130m + 14$

13._____

14. $14q^3 - 9q^2 + q$

14._____

15. $36p^2 + 19p - 6$

15._____

16. $9t^2 + 67t + 28$

16._____

17. $33r^2 - 28r + 4$

17._____

18. $15c^3 + 17c^2 + 4c$

18._____

19. $15b^2 + 34b + 15$

19._____

20. $4n^2 + 35n + 49$

20._____

21. $20m^2 + 104m + 63$

21._____

22. $12y^2 - 23yz + 10z^2$

22._____

23. $3p^2 + 10p + 4$

23._____

Copyright © 2013 Pearson Education, Inc.

24. $45x^3 + 69x^2 + 12x$ 24._____

25. $18y^2 - 49y - 15$ 25._____

26. $12q^2 + q - 6$ 26._____

27. $6p^2 - 13pq - 28q^2$ 27._____

28. $64x^2 - 144x + 72$ 28._____

29. $42m^2 - 85m + 42$ 29._____

30. $240a^4b - 70a^3b + 5a^2b$ 30._____

Name:
Instructor:

Date:
Section:

Chapter 5   FACTORS, DIVISORS, AND FACTORING

5.5   Factoring Binomials

Learning Objectives
A   Factor binomials that are the difference of two squares.
B   Factor the difference of squares, where one or more of the terms is a binomial.
C   Factor the sum and difference of cubes. (Optional)

Getting Ready

Find the following products:

1. $(x+4)(x-4)$

1._____

2. $(2x+5y)(2x-5y)$

2._____

Simplify the following:

3. $(a^3)^3$

3._____

4. $(5x^2)^2$

4._____

Factor the following:

5. $10x^2 + 25y^2$

5._____

Find the following product:

6. $(3x-1)(5x^2+3x+2)$

6._____

Key Terms

Use the vocabulary terms listed below to complete each statement in exercises 1–6.

       perfect squares   **perfect cubes** $(a+b)(a^2-ab+b^2)$
       $(a-b)(a^2+ab+b^2)$ **conjugate pairs**   **binomial**

Copyright © 2013 Pearson Education, Inc.

1. Expressions of the form $a+b$ and $a-b$ are called_____.

2. The numbers 1, 8, 27, 64, 125 are examples of _____.

3. The numbers 1, 4, 9, 16, 25 are examples of _____.

4. A polynomial with two terms is called a(n) _____.

5. $a^3 - b^3 = $ _____.

6. $a^3 + b^3 = $ _____.

Practice Problems

Objective A Factor binomials that are the difference of two squares.

Factor the following completely if possible; if the polynomial is not factorable, write prime.

1. $p^2 - q^2$　　　　　　　　　　　　　　　　　1._____

2. $4a^2 - 9$　　　　　　　　　　　　　　　　　2._____

3. $b^2 + c^2$　　　　　　　　　　　　　　　　　3._____

4. $121x^2 - 144$　　　　　　　　　　　　　　　4._____

5. $49w^2 - 81z^2$　　　　　　　　　　　　　　5._____

6. $169a^2 - 25b^2$　　　　　　　　　　　　　　6._____

7. $100m^2 - 1$　　　　　　　　　　　　　　　7._____

8. $196z^2 - 9$　　　　　　　　　　　　　　　8._____

Name:
Instructor:

Date:
Section:

9. $64p^2 - 81$

9._____

10. $49y^2 - 64$

10._____

11. $9z^2 - 625$

11._____

12. $a^2 - 225$

12._____

13. $25m^4 - n^2$

13._____

14. $100x^4 - 49y^4$

14._____

Factor the following completely.

15. $x^4 - 625$

15._____

16. $y^4 - 16$

16._____

17. $81x^4 - 16$

17._____

18. $256y^4 - 625$

18._____

Objective B Factor the difference of squares, where one or more of the terms is a binomial.

Factor the following completely.

19. $100 - (a+b)^2$

19._____

Copyright © 2013 Pearson Education, Inc.

20. $(x+y)^2 - 121$ 20._____

21. $4(m+n)^2 - 49$ 21._____

22. $25(p-q)^2 - 81$ 22._____

23. $7x^2 - 28$ 23._____

24. $9y^2 - 144$ 24._____

Objective C Factor the sum and difference of cubes. (Optional)

Factor the following sum or difference of cubes.

25. $d^3 + f^3$ 25._____

26. $8a^3 - 1$ 26._____

27. $x^3 + 125$ 27._____

28. $27p^3 + 8$ 28._____

29. $1000m^3 - n^3$ 29._____

30. $24x^3 - 81y^3$ 30._____

31. $343w^3 - 216$ 31._____

32. $216x^3 + 1000$ 32._____

Name:
Instructor:

Date:
Section:

Chapter 5 FACTORS, DIVISORS, AND FACTORING

5.6 Factoring Perfect Square Trinomials

Learning Objectives
A   Recognize perfect square trinomials.
B   Factor perfect square trinomials.

Getting Ready

Square the following:

1. $(x-3y)^2$

1._____

2. $(4a+3b)^2$

2._____

Find the following product:

3. $5(6x)(2y)$

3._____

4. $4(4c)(-3)$

4._____

Factor the following:

5. $ab^3 + 3ab^2 + 10ab$

5._____

Key Terms

Use the vocabulary terms listed below to complete each statement in exercises 1–3.

prime   perfect square trinomial $(a-b)^2$

1. $a^2 - 2ab + b^2 =$ _____.

2. $a^2 - ab + b^2 =$ _____.

3. A _____ results from the square of a binomial.

Copyright © 2013 Pearson Education, Inc.

Practice Problems

Objective A   Recognize perfect square trinomials.

Determine whether the following are perfect square trinomials.

1. $x^2 - 20x + 100$

   1._____

2. $y^2 - 8y - 16$

   2._____

3. $4m^2 + 12m + 9$

   3._____

4. $16k^2 + 20k + 25$

   4._____

Objective B   Factor perfect square trinomials.

Factor each expression, if the trinomial is a perfect square. If it is not a perfect square, write "not a perfect square."

5. $a^2 + 14a + 49$

   5._____

6. $x^2 - 8x + 16$

   6._____

7. $b^2 - 10b + 25$

   7._____

8. $y^2 + 22y + 121$

   8._____

Name: Date:
Instructor: Section:

9. $4t^2 + 12t + 9$

9._____

10. $9p^2 - 30p + 25$

10._____

11. $16q^2 + 56q + 49$

11._____

12. $m^2 + 8m + 64$

12._____

13. $36a^2 - 60ab + 25b^2$

13._____

14. $9y^2 + 24yz + 16z^2$

14._____

15. $3n^2 + 30np + 75p^2$

15._____

16. $4a^3 - 12a^2 + 9a$

16._____

17. $m^4n - 2m^3n^2 + m^2n^3$

17._____

18. $4x^2 + 12xy - 9y^2$

18._____

19. $20x^3y + 60x^2y^2 + 45xy^3$

19._____

20. $11p^2 - 242p + 1331$

20. _____

Name: Date:
Instructor: Section:

Chapter 5 FACTORS, DIVISORS, AND FACTORING

5.7 Mixed Factoring

Learning Objectives
A Factor all types of polynomials.

Getting Ready

Factor the following completely:

1. $a^3 - 9a^2 + 20a$ 1._____

2. $25x^2 - 144$ 2._____

3. $9c^2 + 6c - 3$ 3._____

4. $ax + bx + ay + by$ 4._____

Key Terms

Use the vocabulary terms listed below to complete each statement in exercises 1–5.

| three | prime | grouping | two | greatest common factor |

1. The first step in factoring is to remove the _____ other than 1.

2. If a polynomial has _____ terms try the trial-and-error method or ac method.

3. If a polynomial has _____ terms, check to see if it is a difference of squares or a sum or difference of cubes.

4. If a polynomial has more than three terms try factoring by _____.

5. Factorization is not complete unless each factor is _____.

Practice Problems

Objective A   Factor all types of polynomials.

Factor each of the expressions completely. If the polynomial will not factor, write "prime."

1. $24 - 21x^2$ 1._____

2. $x^2 + 9x + 14$ 2._____

3. $25 - y^2$ 3._____

4. $m^2 + 18m + 81$ 4._____

5. $a^2 + 7a + 49$ 5._____

6. $3p^2 + 8pq + 5q^2$ 6._____

7. $2y^2 - 2y - 112$ 7._____

8. $2b^3 + 15b^2 + 18b$ 8._____

9. $3c^3 - 9c^2 + 6c$ 9._____

Name:
Instructor:

Date:
Section:

10. $a^3b + a^2b - 56ab$

10._____

11. $12p^4q - 6p^3q - 6p^2q$

11._____

12. $t^2 - 121$

12._____

13. $16n^2 - 400$

13._____

14. $2x^2 + 7x + 8$

14._____

15. $2ac + 2bc - 3ad - 3bd$

15._____

16. $36w^2 + 60w + 25$

16._____

17. $30ap + 5bp - 18aq - 3bq$

17._____

18. $5y^2 - 50y + 125$

18._____

19. $mn + mp - 7n - 7p$

19._____

20. $4x^4 - 9y^2$

20._____

21. $18mx - 12nx + 15my - 10ny$ 21._____

22. $64x^2 + 96x + 35$ 22._____

23. $m^2 + 100$ 23._____

24. $p^2 - p - 12$ 24._____

25. $(2a - b)^2 - 4$ 25._____

26. $125x^3 - 64$ 26._____

27. $27y^3 + 216$ 27._____

28. $9m^3 + 9$ 28._____

29. $n^5 + 8n^2$ 29._____

30. $2a^4b^2 - 128ab^2$ 30._____

Name: Date:
Instructor: Section:

Chapter 5   FACTORS, DIVISORS, AND FACTORING

5.8   Solving Quadratic Equations by Factoring

Learning Objectives
A   Solve quadratic equations by factoring.
B   Solve application problems by using quadratic equations.
C   Evaluate quadratic functions.

Getting Ready

Factor the following completely:

1. $p^2 - 9p$ 1._____

2. $4x^2 - 7x + 3$ 2._____

3. $10t^2 - 14t - 12$ 3._____

Solve the following equations:

4. $x - 4 = 0$ 4._____

5. $2x - 5 = 0$ 5._____

Key Terms

Use the vocabulary terms listed below to complete each statement in exercises 1–4.

       parabolas   **zero product**   **quadratic**   **zero**

1. An equation of the form $ax^2 + bx + c = 0$, $a \neq 0$ is called a _____ equation.

2. The graphs of quadratic equations are called _____.

3. The _____ property says if $ab = 0$, then $a = 0$, or $b = 0$, or $a = b = 0$.

4. When solving the equation $ax^2 + bx + c = 0$, $a \neq 0$, factor, then set each factor equal to _____.

Practice Problems

Objective A Solve quadratic equations by factoring.

Solve the following, using the zero product property.

1. $(x-2)(x-11)=0$ 1._____

2. $y(y+13)=0$ 2._____

3. $(2z+7)(5z+3)=0$ 3._____

4. $(a+1)(a-2)(a-8)=0$ 4._____

Solve the following.

5. $w^2-15w+54=0$ 5._____

6. $6p^2+7p-20=0$ 6._____

7. $h^2+10h=0$ 7._____

8. $4q^2=q$ 8._____

9. $72x^2+1=-17x$ 9._____

10. $n^2-12n=-35$ 10._____

11. $a^2-16=0$ 11._____

Name:
Instructor:

Date:
Section:

12. $7y^2 - 28 = 0$

12._____

13. $z(z+4) = 12$

13._____

14. $4x(3x+5) = -7$

14._____

15. $3x^3 + 13x^2 + 4x = 0$

15._____

16. $2a^3 - 22a^2 = -60a$

16._____

17. $x^2 - x = 20$

17._____

18. $4x^3 - 8x^2 - 21x = 0$

18._____

19. $15x^2 - 4 = -7x$

19._____

Objective B Solve application problems by using quadratic equations.

Solve.

20. Find two consecutive positive even integers whose product is 288.

20._____

21. A rectangular blanket is 3 feet longer than it is wide. Find the dimensions of the blanket if the area of the blanket is 54 square feet.

21._____

22. If the difference of twice the square of an integer and five times that integer is 3, find the integer.

22._____

23. If a rock is dropped from the top of a building that is 400 feet high, the equation giving the height of the rock is $h = 400 - 16t^2$, where h is the height in feet above the ground and t is the number of seconds after the rock is dropped. How long will it take the object to reach the ground?

23._____

Objective C Evaluate quadratic functions.

Given $f(x) = x^2 - 2x + 8$, find the following.

24. $f(-5)$

24._____

25. $f(0)$

25._____

26. $f(3)$

26._____

Name:
Instructor:

Date:
Section:

Chapter 6 MULTIPLICATION AND DIVISION OF RATIONAL NUMBERS AND EXPRESSIONS

6.1 Reducing Rational Numbers and Rational Expressions

Learning Objectives
A   Graph rational numbers.
B   Reduce rational numbers to lowest terms.
C   Reduce rational expressions whose numerator and denominator are monomials.

Getting Ready

Write the following whole numbers in terms of their prime factorizations:

1. 54

1._____

2. 156

2._____

Write the following monomials in terms of prime factors:

3. $x^3 y^2$

3._____

4. $a^5 b^4$

4._____

Key Terms

Use the vocabulary terms listed below to complete each statement in exercises 1–4.

    equivalent   lowest terms   rational expression   rational

1. A fraction is reduced to _____ if the numerator and the denominator have no common factors other than 1.

2. Two or more fractions are _____ if they represent the same quantity.

3. A(n) _____ is an algebraic expression of the form $\dfrac{P}{Q}$, where P and Q are polynomials and $Q \neq 0$.

4. A number is _____ if it can be written in the form $\dfrac{a}{b}$ with a and b integers and $b \neq 0$.

Copyright © 2013 Pearson Education, Inc.

Practice Problems

Objective A   Graph rational numbers

Graph the rational numbers. Estimate as needed.

1. $\left\{-1\frac{1}{3},\ -\frac{3}{4},\ 1,\ 2\frac{1}{2}\right\}$

1. _____See graph_____

<-|-|-|-|-|-|-|-|-|-|-|-|-|-|-|-|-|-|-|-|-|-|-|-|-|-|-|->

2. $\left\{-4\frac{5}{6},\ -2\frac{3}{5},\ -\frac{1}{4},\ \frac{1}{2},\ 2\frac{2}{3}\right\}$

2. _____See graph_____

<-|-|-|-|-|-|-|-|-|-|-|-|-|-|-|-|-|-|-|-|-|-|-|-|-|-|-|->

Objective B   Reduce rational numbers to lowest terms.

Reduce the following fractions to lowest terms.

3. $\dfrac{2}{18}$

3. _____

4. $\dfrac{4}{24}$

4. _____

5. $\dfrac{6}{32}$

5. _____

6. $\dfrac{66}{55}$

6. _____

7. $\dfrac{80}{70}$

7. _____

Name:　　　　　　　　　　　　　　　　Date:
Instructor:　　　　　　　　　　　　　　Section:

8. $\dfrac{132}{99}$　　　　　　　　　　　　　　　　8._____

9. $\dfrac{15}{100}$　　　　　　　　　　　　　　　9._____

10. $\dfrac{24}{96}$　　　　　　　　　　　　　　　10._____

11. $\dfrac{810}{700}$　　　　　　　　　　　　　　11._____

12. $\dfrac{65}{78}$　　　　　　　　　　　　　　　12._____

13. $\dfrac{45}{36}$　　　　　　　　　　　　　　　13._____

14. $\dfrac{160}{180}$　　　　　　　　　　　　　　14._____

Objective C　Reduce rational expressions whose numerator and denominator are monomials.

Reduce the following rational expressions to lowest terms.

15. $\dfrac{x^2 y^3}{x^5 y}$　　　　　　　　　　　　　　15._____

16. $\dfrac{a^6 b^{10}}{a^4 b^3}$　　　　　　　　　　　　　　16._____

17. $-\dfrac{p^9 q^3}{p^3 q^7}$　　　　　　　　　　　　　17._____

Copyright © 2013 Pearson Education, Inc.　　　　255

18. $\dfrac{2x^4y^8}{16xy}$ 18._____

19. $\dfrac{-45x^3y^4z^6}{9x^5y^7z}$ 19._____

20. $\dfrac{20m^4n^3}{-5m^5n^8}$ 20._____

21. $\dfrac{15a^6b^4c^2}{75ab^6c^9}$ 21._____

22. $\dfrac{uvw^5}{u^3vw^6}$ 22._____

23. $\dfrac{11x^4y^3}{-33x^2y^2}$ 23._____

24. $-\dfrac{4m^5n^4}{8m^2n^3}$ 24._____

25. $\dfrac{6a^2b^3c}{9a^3b^2c^4}$ 25._____

26. $\dfrac{-32x^6y^5}{16x^2y^3}$ 26._____

Name:
Instructor:
Date:
Section:

Chapter 6 MULTIPLICATION AND DIVISION OF RATIONAL NUMBERS AND EXPRESSIONS

6.2 Further Reduction of Rational Expressions

Learning Objectives
A   Determine the value(s) of the variable(s) for which rational expressions are defined.
B   Reduce rational expressions whose numerators and denominators are not monomials.

Getting Ready

1. Evaluate $\dfrac{7}{0}$.   1._____

2. Solve $3x+5=0$ for x.   2._____

3. Solve $x^2-4x-21=0$ for x.   3._____

4. reduce $\dfrac{56}{72}$ to lowest terms.   4._____

Factor the following polynomials.

5. $2x+6$ 5._____

6. x^2+6x+5 6._____

7. x^2-25 7._____

8. Evaluate $\dfrac{x+3}{x^2-9}$ for $x=4$.   8._____

Key Terms

Use the vocabulary terms listed below to complete each statement in exercises 1–4.

    terms   rational   factor   undefined

1. _____ are separated by + and − signs.

2. _____ implies multiplication.

3. If the denominator of a fraction is 0, then the fraction is _____.

4. A function of the form $f(x) = \dfrac{p(x)}{q(x)}$, $q(x) \neq 0$, where $p(x)$ and $q(x)$ are polynomials is called a(n) _____ function.

Practice Problems

Objective A Determine the value(s) of the variable(s) for which rational expressions are defined.

Find the value(s) of the variable(s) for which the following are defined.

1. $\dfrac{3}{x-4}$

   1._____

2. $\dfrac{6}{y+7}$

   2._____

3. $\dfrac{m-2}{m+1}$

   3._____

4. $\dfrac{2a}{3a+7}$

   4._____

5. $\dfrac{p}{(p+6)(p-7)}$

   5._____

6. $\dfrac{9y}{y^2-81}$

   6._____

7. $\dfrac{x+3}{x^2+8x+15}$

   7._____

8. $\dfrac{6}{6x^2-5x-4}$

   8._____

Name:
Instructor:

Date:
Section:

Objective B Reduce rational expressions whose numerators and denominators are not monomials.

Reduce the following rational expressions to lowest terms.

9. $\dfrac{3(x+2)}{7(x+2)}$

9._____

10. $\dfrac{-4(3p+5)}{11(3p+5)}$

10._____

11. $\dfrac{2x+6}{9x+27}$

11._____

12. $\dfrac{3y-12}{8y-32}$

12._____

13. $\dfrac{-7x-42}{8x+48}$

13._____

14. $\dfrac{-2m+8}{3m-12}$

14._____

15. $\dfrac{x^2+x-2}{x^2+5x+6}$

15._____

Copyright © 2013 Pearson Education, Inc.

16. $\dfrac{y^2-9y+20}{y^2+2y-24}$ 16._____

17. $\dfrac{p^2-9}{p^2-8p+15}$ 17._____

18. $\dfrac{x^2-xy-2y^2}{2x^2+3xy+y^2}$ 18._____

19. $\dfrac{6a^2+5ab+b^2}{12a^2+4ab-b^2}$ 19._____

20. $\dfrac{a-3b}{3b-a}$ 20._____

21. $\dfrac{x^2-49}{7-x}$ 21._____

22. $\dfrac{5-x}{x^2-12x+35}$ 22._____

Name:
Instructor:
Date:
Section:

Chapter 6   MULTIPLICATION AND DIVISION OF RATIONAL NUMBERS AND EXPRESSIONS

6.3   Multiplication of Rational Numbers and Expressions

Learning Objectives
A   Multiply two or more fractions.
B   Multiply mixed numbers and whole numbers with mixed numbers.
C   Multiply rational expressions with monomial numerators and denominators.

Getting Ready

Factor the following:

1. Write 24 in terms of prime factors.

    1._____

2. Write $4\frac{2}{5}$ as an improper fraction.

    2._____

3. Write $12x^3y^5$ in terms of prime factors.

    3._____

4. Find the product: $3a^2b \cdot 10a^3b^4$

    4._____

Key Terms

Use the vocabulary terms listed below to complete each statement in exercises 1–3.

   reduce   **product**   **improper fractions**

1. The _____ of two fractions is found by computing the product of the numerators and dividing by the product of the denominators.

2. With multiplication of fractions you may _____ before actually multiplying.

3. When multiplying mixed numbers convert them to _____ first.

Copyright © 2013 Pearson Education, Inc.

Practice Problems

Objective A Multiply two or more fractions.

Find the products. Express answers reduced to lowest terms.

1. $\dfrac{2}{9} \cdot \dfrac{4}{7}$

    1._____

2. $\dfrac{3}{5} \cdot \left(-\dfrac{1}{8}\right)$

    2._____

3. $\dfrac{7}{10} \cdot \dfrac{3}{5}$

    3._____

4. $\left(-\dfrac{3}{4}\right)\left(-\dfrac{5}{7}\right)$

    4._____

5. $\dfrac{3}{10} \cdot \dfrac{5}{6}$

    5._____

6. $18 \cdot \left(-\dfrac{7}{3}\right)$

    6._____

7. $\dfrac{3}{8} \cdot \dfrac{16}{9} \cdot \dfrac{36}{5}$

    7._____

8. $\dfrac{12}{30} \cdot \dfrac{25}{24}$

    8._____

Name: Date:
Instructor: Section:

9. $\dfrac{9}{20} \cdot \dfrac{5}{18}$

9._____

10. $\dfrac{5}{14} \cdot \dfrac{28}{3} \cdot \dfrac{15}{49}$

10._____

Objective B Multiply mixed numbers and whole numbers with mixed numbers.

Find the products. Express answers reduced to lowest terms.

11. $2\dfrac{1}{5} \cdot 10$

11._____

12. $5\dfrac{1}{5} \cdot 2\dfrac{1}{3}$

12._____

13. $-4\dfrac{3}{5} \cdot \left(-1\dfrac{2}{15}\right)$

13._____

14. $\dfrac{7}{8} \cdot 5\dfrac{3}{4}$

14._____

15. $-\dfrac{16}{9} \cdot 4\dfrac{3}{8}$

15._____

16. $24 \cdot 6\dfrac{7}{12}$

16._____

17. $\left(-6\dfrac{3}{16}\right) \cdot \dfrac{8}{7} \cdot \dfrac{4}{9}$

17._____

18. $8 \cdot 3\dfrac{2}{3} \cdot 12\dfrac{1}{2}$

18._____

19. $5\dfrac{3}{10} \cdot (-20)$

19._____

20. $\left(-5\dfrac{2}{7}\right)\left(-2\dfrac{1}{10}\right)$

20._____

Objective C Multiply rational expressions with monomial numerators and denominators.

Find the products. Express answers reduced to lowest terms and assume that all variables have nonzero values.

21. $\dfrac{x^5}{y^7} \cdot \dfrac{y^2}{x^3}$

21._____

22. $\dfrac{a^2}{b^5} \cdot \dfrac{b^3}{a}$

22._____

23. $\dfrac{p^2 q^2}{p^4 q^3} \cdot \dfrac{pq}{p^3 q^4}$

23._____

24. $\dfrac{2m^2 n}{3mn^2} \cdot \dfrac{6m^3 n}{5m^4 n^2}$

24._____

Name: Date:
Instructor: Section:

25. $\dfrac{8c^3}{9d^2} \cdot \dfrac{3d^4}{4c^2}$ 25._____

26. $\dfrac{14x^4y^5}{15x^3y} \cdot \dfrac{3x^6y^3}{7x^2y}$ 26._____

27. $\dfrac{18a^2b^3c^4}{33a^4b^4c^2} \cdot \dfrac{22abc^6}{9a^4b^2c^3}$ 27._____

28. $\dfrac{8u^2v^5}{3uv^4} \cdot \dfrac{6u^3v}{4u^2v^2}$ 28._____

Answer the following.

29. It takes $2\dfrac{3}{8}$ feet of ribbon for a Christmas wreath. 29._____
How many feet of ribbon would it take for 16 of these wreaths?

30. A running track is $2\frac{1}{4}$ miles long. Joey has run $\frac{2}{3}$ of the way around the track. How far has Joey run?

30._____

Name:
Instructor:

Date:
Section:

Chapter 6 MULTIPLICATION AND DIVISION OF RATIONAL NUMBERS AND EXPRESSIONS

6.4 Further Multiplication of Rational Expressions

Learning Objectives
A Multiply rational expressions whose numerators and denominators are not monomials.
B Multiply rational expressions that contain factors of the form $a-b$ and $b-a$.

Getting Ready

Multiply the following rational numbers and rational expressions with monomial numerators and denominators:

1. $\dfrac{4}{16} \cdot \dfrac{12}{8}$

   1._____

2. $\dfrac{xy}{x^2 y^2} \cdot \dfrac{x^3 y^2}{x^2 y^3}$

   2._____

Factor the following polynomials:

3. $6x - 18$

   3._____

4. $x^2 + 2x - 35$

   4._____

5. $x^2 - 81$

   5._____

Key Terms

Use the vocabulary terms listed below to complete each statement in exercises 1–3.

-1 divide   factor

1. When multiplying rational expressions one of the first steps in the procedure is to_____ the numerators and denominators into prime factors.

2. Another step in the procedure of multiplying rational expressions is to _____ the numerator and denominator by factors common to both.

3. $\dfrac{a-b}{b-a} = $ _____.

Practice Problems

Objective A Multiply rational expressions whose numerators and denominators are not monomials.

Find the products. Express answers reduced to lowest terms.

1. $\dfrac{(x+y)^3}{12w} \cdot \dfrac{18w^2}{x+y}$

    1._____

2. $\dfrac{4a^2b^6}{9(v+w)} \cdot \dfrac{3(v+w)^2}{24ab^3}$

    2._____

3. $\dfrac{3x+6}{4x-16} \cdot \dfrac{2x-8}{6x+12}$

    3._____

4. $\dfrac{3x-3}{5x+10} \cdot \dfrac{4x+8}{7x-7}$

    4._____

5. $\dfrac{x^2-16}{x-4} \cdot \dfrac{x-3}{x^2-9}$

    5._____

6. $\dfrac{25x^2-1}{2x+3} \cdot \dfrac{4x^2-9}{5x-1}$

    6._____

Name:
Instructor:

Date:
Section:

7. $\dfrac{2a+5}{6a-1} \cdot \dfrac{6a+1}{4a^2-25}$

7._____

8. $\dfrac{8x^2y^4}{3x-9} \cdot \dfrac{6x-18}{4xy^3}$

8._____

9. $\dfrac{x^2-3x-4}{x^2-2x-8} \cdot \dfrac{x^2+5x+6}{x^2+6x+5}$

9._____

10. $\dfrac{2a^2-a-15}{3a^2-8a-3} \cdot \dfrac{6a^2+37a-35}{2a^2+19a+35}$

10._____

11. $\dfrac{6x^2-23x+20}{6x^2+11x+4} \cdot \dfrac{8x^2-10x-7}{8x^2-34x+35}$

11._____

12. $\dfrac{x^2-100}{x^2-13x+30} \cdot \dfrac{x^2+8x-33}{x^2-121}$

12._____

Objective B Multiply rational expressions that contain factors of the form $a-b$ and $b-a$.

Find the products. Express answers reduced to lowest terms.

13. $\dfrac{b-7}{6} \cdot \dfrac{12}{7-b}$

13._____

Copyright © 2013 Pearson Education, Inc.

14. $\dfrac{p-9}{15} \cdot \dfrac{5}{9-p}$

14. _____

15. $\dfrac{x^2-4}{x+3} \cdot \dfrac{x^2+6x+9}{2-x}$

15. _____

16. $\dfrac{q^2-25}{q^2+9q+20} \cdot \dfrac{q^2+12q+32}{5-q}$

16. _____

17. $\dfrac{6x^2-x-1}{2x^2-13x-24} \cdot \dfrac{8x^2+22x+15}{1+x-6x^2}$

17. _____

18. $\dfrac{9x^2+71x-8}{27x^2+42x-5} \cdot \dfrac{3x^2+26x+35}{x^2+15x+56}$

18. _____

19. $\dfrac{y^2-13y+42}{y^2-8y+15} \cdot \dfrac{y-5}{7-y}$

19. _____

20. $\dfrac{x^2+1}{x^2-x-12} \cdot \dfrac{4-x}{x+1}$

20. _____

Name: Date:
Instructor: Section:

Chapter 6 MULTIPLICATION AND DIVISION OF RATIONAL NUMBERS AND EXPRESSIONS

6.5 Division of Rational Numbers and Expressions

Learning Objectives
A Divide rational numbers.
B Solve application problems with rational numbers.
C Divide rational expressions.

Getting Ready

1. Convert $6\frac{2}{3}$ into an improper fraction.   1._____

Factor each of the following:

2. 18   2._____

3. $x^2 - 8x$ 3._____

4. $2x^2 - 9x - 5$ 4._____

5. $x^2 - 25$ 5._____

Multiply the following rational numbers and expressions:

6. $\dfrac{3}{8} \cdot \dfrac{2}{15}$ 6._____

7. $\dfrac{a^3 b^2}{a^4 b^3} \cdot \dfrac{a^4 b^2}{a^2 b}$ 7._____

8. $\dfrac{2x^2 + 9x + 4}{x^2 - 16} \cdot \dfrac{x^2 - 4x}{2x^2 + 11x + 5}$ 8._____

Copyright © 2013 Pearson Education, Inc.   271

Key Terms

Use the vocabulary terms listed below to complete each statement in exercises 1–4.

    divide   **expressions**   **reciprocal**   **multiplicative inverses**

1. Another name for multiplicative inverse is _____.

2. Two numbers are _____ if their product is 1.

3. To _____ by a fraction, multiply by its reciprocal.

4. Division of rational _____ is done in the same manner as division of rational numbers.

Practice Problems

Objective A Divide rational numbers.

Find the quotients of fractions. Express answers reduced to lowest terms.

1. $\dfrac{3}{4} \div \dfrac{9}{8}$ 1._____

2. $-\dfrac{2}{9} \div \dfrac{5}{9}$ 2._____

3. $-\dfrac{16}{15} \div \left(-\dfrac{4}{3}\right)$ 3._____

4. $\dfrac{32}{45} \div \left(-\dfrac{8}{9}\right)$ 4._____

5. $-36 \div \dfrac{21}{4}$ 5._____

6. $2\dfrac{1}{4} \div \dfrac{15}{16}$ 6._____

Name:
Instructor:

Date:
Section:

7. $-5\dfrac{5}{6} \div \dfrac{11}{12}$

7._____

8. $-4\dfrac{3}{8} \div \left(-2\dfrac{4}{9}\right)$

8._____

9. $15 \div 7\dfrac{20}{35}$

9._____

10. $6\dfrac{7}{12} \div \left(-3\dfrac{3}{4}\right)$

10._____

11. $8\dfrac{4}{9} \div 6$

11._____

12. $10 \div \left(-3\dfrac{2}{5}\right)$

12._____

Objective B Solve application problems with rational numbers.

Solve the following.

13. If $4\dfrac{1}{2}$ pounds of steak cost $54, what is the cost of 1 pound?

13._____

14. A bag of candy that is $\dfrac{3}{4}$ full is divided equally among 6 people. What part of a full bag did each receive?

14._____

Objective C Divide rational expressions.

Find the quotients of the given rational expressions. Express answers reduced to lowest terms.

15. $\dfrac{x^4}{y^5} \div \dfrac{x^3}{y^8}$

15._____

16. $\dfrac{a^5 b^7}{c^4 d^3} \div \dfrac{ab^4}{c^2 d}$

16._____

17. $\dfrac{5p^2 q^8}{4x^5 y} \div \dfrac{45 p^3 q^4}{8x^2 y^3}$

17._____

18. $\dfrac{2a-6}{15a} \div \dfrac{5a-15}{3a^3}$

18._____

19. $\dfrac{6x-42}{21y^2} \div \dfrac{7-x}{7x-28}$

19._____

Name: Date:
Instructor: Section:

20. $\dfrac{x^2+5x+6}{x^2-4x-21} \div \dfrac{7+x}{2x^2-13x-7}$ **20.** _____

21. $\dfrac{p^2-81}{p^2+12p+20} \div \dfrac{9-p}{p+10}$ **21.** _____

22. $\dfrac{9n^2-9n-10}{6n^2-17n-14} \div \dfrac{5n^2-29n-6}{10n^2-33n-7}$ **22.** _____

23. $\dfrac{y+5}{y-5} \div \dfrac{y^2+10y+25}{y^2-10y+25}$ **23.** _____

24. $\dfrac{w^2-7w+6}{w^2+8w+15} \div \dfrac{w^2-5w-6}{w^2+14w+33}$ **24.** _____

25. $\dfrac{49-a^2}{a^2+8a+7} \div \dfrac{a^2-10a+21}{a^2+3a+2}$

25. _____

Name:
Instructor:
Date:
Section:

Chapter 6 MULTIPLICATION AND DIVISION OF RATIONAL NUMBERS AND EXPRESSIONS

6.6 Division of Polynomials (Long Division)

Learning Objectives
A Divide two polynomials, using long division.

Getting Ready

1. Divide $16\overline{)384}$.

1._____

Simplify the following, using $a^m \cdot a^n = a^{m+n}$ *and* $\dfrac{a^m}{a^n} = a^{m-n}$:

2. $3a^2 \cdot 5a^3$

2._____

3. $\dfrac{9a^4}{3a}$

3._____

Add the following like terms:

4. $6x - 7x$

4._____

5. $12x - 8x$

5._____

Key Terms

Use the vocabulary terms listed below to complete each statement in exercises 1–3.

    zero   **quotient**   **descending powers**

1. The word _____ indicates division.

2. Before performing long division, write the divisor and the dividend in _____ of the variable.

3. If there is a missing power of the variable, _____ times the variable may be inserted to hold the place.

Practice Problems

Objective A   Divide two polynomials, using long division.

Find the following quotients, using long division.

1. $\dfrac{x^2 + 12x + 35}{x + 7}$

    1._____

2. $\dfrac{x^2 + 4x - 32}{x - 4}$

    2._____

3. $\dfrac{6x^2 - 5x - 4}{2x + 1}$

    3._____

4. $\dfrac{14x^2 + 25x + 6}{2x + 3}$

    4._____

5. $\dfrac{2x^3 + 7x^2 - 7x - 30}{x - 2}$

    5._____

Name:
Instructor:

Date:
Section:

6. $\dfrac{3x^3 - 35x^2 + 93x + 35}{3x + 1}$

6._____

7. $\dfrac{x^3 + 125}{x + 5}$

7._____

8. $\dfrac{x^3 - 15x^2 + 62x - 70}{x - 9}$

8._____

9. $\dfrac{x^4 - 10000}{x + 10}$

9._____

10. $\dfrac{x^3 - 16}{x - 2}$

10._____

11. $\dfrac{x^4 + x^3 - 32x^2 + 12x + 149}{x - 3}$ 11._____

12. $\dfrac{16x + 3x^3 - 5}{x - 1}$ 12._____

13. $\dfrac{2x^2 - 47x + 24x^3 - 15}{2x - 3}$ 13._____

14. $\dfrac{x^3 - 24}{x - 4}$ 14._____

15. $\dfrac{8x^4 + 24x^3 + 18x^2 + 16x + 30}{2x + 3}$ 15._____

Name:
Instructor:

Date:
Section:

Chapter 7 ADDITION AND SUBTRACTION OF RATIONAL NUMBERS AND EXPRESSIONS

7.1 Addition and Subtraction of Rational Numbers and Expressions with Like Denominators

Learning Objectives
A   Add and subtract fractions with common denominators.
B   Add and subtract mixed numbers whose fractional parts have common denominators.
C   Add and subtract rational expressions with common denominators.

Getting Ready

Add or subtract the following integers:

1. $13-(-8)$ 1._____

2. $15-6+9$ 2._____

3. $21-4-13$ 3._____

4. Convert $5\frac{3}{7}$ to an improper fraction.   4._____

Factor the following whole number and polynomial:

5. 28 5._____

6. $x^2-11x+28$ 6._____

Add or subtract the following polynomials:

7. $(x^2+5x)+(4x-7)$ 7._____

8. $(x^2+3x)-(5x+21)$ 8._____

Reduce the following to lowest terms:

9. $\dfrac{14}{35}$ 9._____

Copyright © 2013 Pearson Education, Inc.   281

10. $\dfrac{3x-12}{x-4}$

10._____

11. $\dfrac{x^2+7x+10}{x^2+8x+15}$

11._____

Key Terms

Use the vocabulary terms listed below to complete each statement in exercises 1–4.

difference   **numerators**   **common denominators**   **improper fractions**

1. To add or subtract fractions they must have _____.

2. To add fractions with common denominators, add the _____ and put the sum over the common denominator.

3. To subtract fractions with common denominators, put the _____ of the numerators over the common denominator.

4. Mixed numbers do not have to be converted to _____ in order to add or subtract them.

Practice Problems

Objective A   Add and subtract fractions with common denominators.

Find the following sums or differences. Express the answers reduced to lowest terms.

1. $\dfrac{7}{10}+\dfrac{1}{10}$

1._____

2. $\dfrac{3}{5}-\dfrac{1}{5}$

2._____

3. $\dfrac{7}{12}-\dfrac{5}{12}$

3._____

4. $\dfrac{5}{9}+\dfrac{2}{9}$

4._____

5. $\dfrac{10}{21}+\dfrac{5}{21}$

5._____

Name:
Instructor:

Date:
Section:

6. $\dfrac{5}{18} - \dfrac{7}{18}$

6._____

7. $\dfrac{5}{6} - \dfrac{-7}{6}$

7._____

8. $\dfrac{4}{13} - \dfrac{9}{13}$

8._____

9. $\dfrac{7}{15} - \dfrac{8}{15} + \dfrac{2}{15}$

9._____

10. $-\dfrac{11}{30} - \dfrac{-17}{30} + \dfrac{13}{30}$

10._____

Objective B Add and subtract mixed numbers whose fractional parts have common denominators.

Add or subtract the following mixed numbers. Express the answers reduced to lowest terms.

11. $2\dfrac{5}{9} + 6\dfrac{1}{9}$

11._____

12. $3\dfrac{2}{5} - 8\dfrac{4}{5}$

12._____

13. $9\dfrac{7}{8} - 10\dfrac{1}{8}$

13._____

14. $16\dfrac{15}{31} + 3\dfrac{6}{31} - 9\dfrac{19}{31}$

14._____

Objective C Add and subtract rational expressions with common denominators.

Find the sum or difference of each of the rational expressions. Express your answers reduced to lowest terms.

15. $\dfrac{2}{x} + \dfrac{5}{x}$

15._____

16. $\dfrac{7}{p} - \dfrac{q}{p}$

16._____

Copyright © 2013 Pearson Education, Inc.

17. $\dfrac{9y}{2z^2} + \dfrac{7y}{2z^2}$ 17._____

18. $\dfrac{5}{3m} + \dfrac{1}{3m} - \dfrac{16}{3m}$ 18._____

19. $\dfrac{2}{x+3} + \dfrac{5}{x+3}$ 19._____

20. $\dfrac{8}{z-4} - \dfrac{10}{z-4}$ 20._____

21. $\dfrac{8n}{2-m} - \dfrac{16n}{2-m}$ 21._____

22. $\dfrac{5x}{x+9} + \dfrac{4x}{x+9} - \dfrac{7x}{x+9}$ 22._____

23. $\dfrac{2}{v+8} + \dfrac{6}{v+8} - \dfrac{v}{v+8}$ 23._____

24. $\dfrac{3b}{2b+9} - \dfrac{5b}{2b+9} + \dfrac{36}{2b+9}$ 24._____

25. $\dfrac{x^2}{x^2+7x+6} + \dfrac{3x-18}{x^2+7x+6}$ 25._____

26. $\dfrac{y^2}{y^2-5y+6} - \dfrac{12y-27}{y^2-5y+6}$ 26._____

27. $\dfrac{2a^2+48}{a^2-36} - \dfrac{a^2-13a+6}{a^2-36}$ 27._____

28. $\dfrac{2}{9} + \dfrac{11}{-9}$ 28._____

29. $\dfrac{-2}{u-3} - \dfrac{5}{3-u}$ 29._____

30. $\dfrac{7x+8y}{4x-3y} + \dfrac{5x+11y}{3y-4x}$ 30._____

Name: Date:
Instructor: Section:

Chapter 7 ADDITION AND SUBTRACTION OF RATIONAL NUMBERS AND EXPRESSIONS

7.2 Least Common Multiple and Equivalent Rational Expressions

Learning Objectives
A   Find the least common multiple of two or more integers.
B   Find the least common multiple of two or more polynomials.
C   Change a fraction into an equivalent fraction with a specified denominator.
D   Change a rational expression into an equivalent rational expression with a specified denominator.

Getting Ready

1. Write 60 in terms of its prime factors. 1._____

Factor the following:

2. $9p^2 - 16$ 2._____

3. $10x^2 - 13x - 3$ 3._____

Divide the following whole numbers and monomials:

4. $\dfrac{112}{14}$ 4._____

5. $\dfrac{35y^5}{7y^3}$ 5._____

Multiply the following monomials:

6. $6x^3 \cdot 4x$ 6._____

7. $4x^2 y \cdot 3x^3 y^4$ 7._____

Multiply the following rational numbers and expressions. Do not reduce the product to lowest terms.

8. $\dfrac{5}{5} \cdot \dfrac{3}{7}$ 8._____

Copyright © 2013 Pearson Education, Inc.

9. $\dfrac{3m^3}{3m^3} \cdot \dfrac{14n}{6m^3}$

9._____

10. $\dfrac{y+2}{y+2} \cdot \dfrac{5}{y(y-1)}$

10._____

Key Terms

Use the vocabulary terms listed below to complete each statement in exercises 1–4.

Fundamental Property of Fractions   equivalent
least common multiple   GCF

1. The _____ of two integers is the smallest positive integer that is divisible by the two integers.

2. The LCM is the reverse of a(n) _____.

3. Two or more fractions are _____ if they represent the same quantity.

4. The _____ states if $\dfrac{a}{b}$ is a fraction and c is any number except 0, then $\dfrac{a}{b} = \dfrac{a}{b} \cdot \dfrac{c}{c} = \dfrac{ac}{bc}$.

Practice Problems

Objective A **Find the least common multiple of two or more integers.**

Find the LCM of the following whole numbers.

1. 8 and 9

1._____

2. 16 and 24

2._____

3. 18 and 54

3._____

4. 10, 20, and 45

4._____

Name:
Instructor:
Date:
Section:

Objective B Find the least common multiple of two or more polynomials.

Find the LCM of the following polynomials.

5. x^2y^3 and x^6y^2

5._____

6. ab^5 and a^2b^3

6._____

7. $2p^4q^2$ and $8p^3q$

7._____

8. $t+4$ and $t-1$

8._____

9. $m+4$ and $5m+20$

9._____

10. $x-7$ and x^2+x-56

10._____

11. y^2-100 and $y^2+13y+30$

11._____

12. x^2-81, $x^2-10x+9$, and $x^2+11x+18$

12._____

Objective C Change a fraction into an equivalent fraction with a specified denominator.

Find each missing numerator so that the fractions are equal.

13. $\dfrac{2}{3}=\dfrac{?}{36}$

13._____

14. $2=\dfrac{?}{7}$

14._____

15. $\dfrac{c}{d}=\dfrac{?}{8d}$

15._____

16. $\dfrac{5}{6y}=\dfrac{?}{42y}$

16._____

Copyright © 2013 Pearson Education, Inc.

17. $\dfrac{8x}{11y} = \dfrac{?}{22y^3}$

17._____

18. $\dfrac{16b^4}{7a^2} = \dfrac{?}{21a^6}$

18._____

19. $\dfrac{2m^3}{9pq^4} = \dfrac{?}{45p^3q^7}$

19._____

Objective D Change a rational expression into an equivalent rational expression with a specified denominator.

Find each missing numerator so that the rational expressions are equal.

20. $\dfrac{x}{x-4} = \dfrac{?}{3x-12}$

20._____

21. $\dfrac{y}{3y+9} = \dfrac{?}{12y+36}$

21._____

22. $\dfrac{8p}{12-5p} = \dfrac{?}{48-20p}$

22._____

23. $\dfrac{7}{a+3} = \dfrac{?}{a^2-4a-21}$

23._____

24. $\dfrac{3v}{6+v} = \dfrac{?}{48+2v-v^2}$

24._____

25. $\dfrac{9y}{y^2+4y} = \dfrac{?}{y(y+4)(y-3)}$

25._____

26. $\dfrac{n+1}{n^2-49} = \dfrac{?}{(n+7)(n-2)(n-7)}$

26._____

27. $\dfrac{a+2}{a^2+8a+16} = \dfrac{?}{(a+4)(a-1)(a+4)}$

27._____

28. $\dfrac{x-10}{x^2-15x+56} = \dfrac{?}{x(x-8)(x-7)}$

28._____

Name: Date:
Instructor: Section:

Chapter 7 ADDITION AND SUBTRACTION OF RATIONAL NUMBERS AND EXPRESSIONS

7.3 The Least Common Denominator of Fractions and Rational Expressions

Learning Objectives
A Find the least common denominator for two or more fractions and convert each fraction into an equivalent fraction with the least common denominator as its denominator.
B Find the least common denominator for two or more rational expressions and convert each rational expression into an equivalent rational expression with the least common denominator as its denominator.

Getting Ready

Find the least common multiple for each of the following:

1. 15 and 25

1._____

2. $8xy^3$ and $12x^2y$

2._____

3. $x^2 - 2x - 3$ and $x^2 + x - 12$

3._____

4. Multiply $6x \cdot 3x^2 y^5$

4._____

5. Factor $x^2 - 4x - 32$.

5._____

Multiply the rational numbers and expressions. Do not reduce the product.

6. $\dfrac{3}{3} \cdot \dfrac{3}{8}$

6._____

7. $\dfrac{2x}{2x} \cdot \dfrac{7}{3x^3 y^4}$

7._____

8. $\dfrac{x-3}{x-3} \cdot \dfrac{x+5}{(x+1)(x+4)}$

8._____

Copyright © 2013 Pearson Education, Inc.   289

Key Terms

Use the vocabulary terms listed below to complete each statement in exercises 1–2.

least common denominator   **rational numbers**

1. The _____ of two or more fractions is the least common multiple of the denominators.

2. Writing rational expressions with the least common denominator follows the same procedure as writing _____ with the least common denominator.

Practice Problems

Objective A Find the least common denominator for two or more fractions and convert each fraction into an equivalent fraction with the least common denominator as its denominator.

Find the least common denominator. Then write each fraction as an equivalent fraction with the LCD as its denominator.

1. $\dfrac{2}{9}$ and $\dfrac{5}{8}$ 1._____

2. $\dfrac{5}{12}$ and $\dfrac{11}{24}$ 2._____

3. $\dfrac{3}{8}$ and $\dfrac{5}{18}$ 3._____

4. $\dfrac{9}{20}$ and $\dfrac{17}{30}$ 4._____

5. $\dfrac{5}{54}$ and $\dfrac{35}{36}$ 5._____

6. $\dfrac{1}{6}, \dfrac{5}{24},$ and $\dfrac{11}{36}$ 6._____

7. $\dfrac{3}{4}, \dfrac{9}{20},$ and $\dfrac{21}{35}$ 7._____

8. $\dfrac{1}{6}, \dfrac{1}{9},$ and $\dfrac{1}{4}$ 8._____

Name:
Instructor:

Date:
Section:

Objective B Find the least common denominator for two or more rational expressions and convert each rational expression into an equivalent rational expression with the least common denominator as its denominator.

Find the least common denominator. Then write each rational expression as an equivalent expression with the LCD as its denominator.

9. $\dfrac{5}{a}$ and $\dfrac{4}{b}$

9._____

10. $\dfrac{7}{m}$ and $\dfrac{6}{n}$

10._____

11. $\dfrac{p}{q}$ and $\dfrac{u}{v}$

11._____

12. $\dfrac{7}{ab}$ and $\dfrac{3}{bc}$

12._____

13. $\dfrac{9}{xy}$ and $\dfrac{20}{yz}$

13._____

14. $\dfrac{2}{m^2 n}$ and $\dfrac{4}{mn^3}$

14._____

15. $\dfrac{8a}{x^2 yz^3}$ and $\dfrac{7b}{xy^4 z^2}$

15._____

16. $\dfrac{m}{n^2 p^4}$ and $\dfrac{2}{n^3 p^2}$

16._____

17. $\dfrac{4a}{11b^3 c}$ and $\dfrac{5t}{22b^4 c^3}$

17._____

18. $\dfrac{x}{8m^2 n^3}$ and $\dfrac{y}{12m^3 n^2}$

18._____

Copyright © 2013 Pearson Education, Inc.

19. $\dfrac{1}{c+1}$ and $\dfrac{2}{c+3}$

19._____

20. $\dfrac{7}{x+9}$ and $\dfrac{5}{x+8}$

20._____

21. $\dfrac{m}{4m-12}$ and $\dfrac{2m}{7m-21}$

21._____

22. $\dfrac{y}{6y-18}$ and $\dfrac{5y}{4y-12}$

22._____

23. $\dfrac{x}{x^2-64}$ and $\dfrac{4x}{x^2-15x+56}$

23._____

24. $\dfrac{2t}{t^2-121}$ and $\dfrac{3t}{t^2-12t+11}$

24._____

25. $\dfrac{w+6}{w^2-w-6}$ and $\dfrac{w-1}{w^2-7w+12}$

25._____

26. $\dfrac{p-4}{p^2+3p+2}$ and $\dfrac{3-p}{p^2+4p+3}$

26._____

27. $\dfrac{x+1}{x^2-9x+14}$ and $\dfrac{x+8}{x^2+7x-18}$

27._____

28. $\dfrac{y-5}{y^2-10y+24}$ and $\dfrac{y-10}{y^2-12y+32}$

28._____

29. $\dfrac{n+2}{n^3+n^2-20n}$ and $\dfrac{n-6}{n^3+5n^2}$

29._____

30. $\dfrac{a+1}{a^5-4a^3}$ and $\dfrac{a+7}{a^3+3a^2+2a}$

30._____

Name: Date:
Instructor: Section:

Chapter 7 ADDITION AND SUBTRACTION OF RATIONAL NUMBERS AND EXPRESSIONS

7.4 Addition and Subtraction of Rational Numbers and Expressions with Unlike Denominators

Learning Objectives
A   Add fractions and mixed numbers with unlike denominators.
B   Add rational expressions with unlike denominators.

Getting Ready

1. Write $3\frac{5}{8}$ as an improper fraction.   1._____

Add the following fractions and rational expressions:

2. $\dfrac{6}{45}+\dfrac{13}{45}$ 2._____

3. $\dfrac{3x+8}{6x}+\dfrac{3x-7}{6x}$ 3._____

4. $\dfrac{4x-8}{5(x-3)}-\dfrac{2x-2}{5(x-3)}$ 4._____

Factor the following:

5. $2x^2-10x-28$ 5._____

6. $a^2-14a+49$ 6._____

Rewrite each of the following as equivalent fractions with a common denominator:

7. $\dfrac{3}{8}$ and $\dfrac{5}{7}$ 7._____

8. $\dfrac{3}{5m}$ and $\dfrac{7}{9m}$ 8._____

Copyright © 2013 Pearson Education, Inc.   293

9. $\dfrac{x+4}{3x-9}$ and $\dfrac{x+5}{2x-6}$

9._____

Key Terms

Use the vocabulary terms listed below to complete each statement in exercises 1–2.

trinomials   **least common denominator**

1. To add fractions with unlike denominators, change each fraction to an equivalent fraction with the _____ as its denominator.

2. If the denominators are binomials or _____, it is often necessary to factor the denominators to find the LCD.

Practice Problems

Objective A Add fractions and mixed numbers with unlike denominators.

Find each sum or difference. Express the answers as fractions reduced to lowest terms.

1. $\dfrac{3}{8}+\dfrac{7}{9}$

1._____

2. $\dfrac{2}{3}-\dfrac{1}{5}$

2._____

3. $\dfrac{6}{5}-14$

3._____

4. $12-\dfrac{3}{7}$

4._____

5. $\dfrac{3}{14}+\dfrac{9}{7}$

5._____

Name:
Instructor:

Date:
Section:

6. $\dfrac{11}{17} - \dfrac{21}{34}$

6._____

7. $\dfrac{7}{10} + 5$

7._____

8. $\dfrac{17}{20} - \dfrac{13}{15}$

8._____

9. $2\dfrac{3}{7} + 8\dfrac{1}{21}$

9._____

10. $4\dfrac{5}{9} - 2\dfrac{14}{15}$

10._____

11. $\dfrac{4}{5} + \dfrac{17}{20} - \dfrac{8}{15}$

11._____

12. $\dfrac{11}{6} - \dfrac{5}{4} - \dfrac{3}{10}$

12._____

13. $3\dfrac{7}{9} + 15\dfrac{5}{12}$

13._____

14. $29\dfrac{1}{4} - 15\dfrac{1}{8}$

14._____

15. $\frac{1}{2}+7-4\frac{5}{8}$

15._____

16. $-12-2\frac{1}{3}+\frac{5}{6}$

16._____

Objective B Add rational expressions with unlike denominators.

Find each sum or difference. Express the answers as rational expressions reduced to lowest terms.

17. $\frac{5}{x}-\frac{x}{y}$

17._____

18. $\frac{3}{m}+\frac{2}{n}$

18._____

19. $\frac{7}{9t}-\frac{13}{18t}$

19._____

20. $\frac{2a}{5b}+\frac{6a}{15b}$

20._____

21. $\frac{p+5}{4p}-\frac{p-3}{10p}$

21._____

Name:
Instructor:

Date:
Section:

22. $\dfrac{8n-3}{14p^3m^2} + \dfrac{4n+5}{21p^2m^4}$

22._____

23. $\dfrac{6}{r-5} - \dfrac{10}{r}$

23._____

24. $\dfrac{6}{w+2} + \dfrac{5}{w-1}$

24._____

25. $\dfrac{9}{x-7} - \dfrac{8}{x-6}$

25._____

26. $\dfrac{x+3}{x-7} + \dfrac{x-2}{x+6}$

26._____

27. $\dfrac{2y+1}{3y-1} + \dfrac{8y+3}{2y+5}$

27._____

Copyright © 2013 Pearson Education, Inc.

28. $\dfrac{4w+3}{2w+7} - \dfrac{w-1}{5w-6}$

28._____

29. $\dfrac{4t}{t^2+5t+6} + \dfrac{5t}{t^2-5t-14}$

29._____

30. $\dfrac{x^2+4x}{x^2+8x+16} + \dfrac{5}{x+4}$

30._____

31. $\dfrac{y-2}{y^2+11y+24} + \dfrac{y+8}{y^2+7y+12}$

31._____

32. $\dfrac{x+4}{x^2+6x+5} + \dfrac{5x-15}{x^2+2x-15}$

32._____

33. $\dfrac{w-5}{w^2+10w+25} - \dfrac{w+8}{w^2+4w-5}$

33._____

Name:
Instructor:

Date:
Section:

Chapter 7 ADDITION AND SUBTRACTION OF RATIONAL NUMBERS AND EXPRESSIONS

7.5 Complex Fractions

Learning Objectives
A Simplify complex fractions whose numerators and/or denominators are rational numbers.
B Simplify complex fractions whose numerators and/or denominators are rational expressions.

Getting Ready

Divide the following rational numbers and expressions:

1. $\dfrac{3}{4} \div \dfrac{7}{8}$

   1._____

2. $\dfrac{xy^2}{z} \div \dfrac{x^3 y}{z^3}$

   2._____

Reduce the following to lowest terms:

3. $\dfrac{28xy^2 z}{7x^3 yz^5}$

   3._____

4. $\dfrac{y^2 - 16}{y^2 - 4y}$

   4._____

5. Factor $x^2 + 15x + 54$.

   5._____

6. Find the LCD for 1, $\dfrac{9}{a}$, and $\dfrac{16}{a^2}$.

   6._____

Simplify the following, using the distributive property:

7. $b^3 \left(1 + \dfrac{4}{b}\right)$

   7._____

Copyright © 2013 Pearson Education, Inc.

8. $(y-7)\left(y+\dfrac{4}{y-7}\right)$

8._____

Multiply the following rational numbers and expressions:

9. $\dfrac{13}{8} \cdot \dfrac{6}{19}$

9._____

10. $\dfrac{a^2 b^3}{c^2} \cdot \dfrac{c^5}{ab^4}$

10._____

Key Terms

Use the vocabulary terms listed below to complete each statement in exercises 1–3.

complex	simplify	one

1. To _____ a complex fraction means to write it as an ordinary fraction in the form of $\dfrac{a}{b}$, where a and b are integers and $b \neq 0$.

2. A(n)_____ fraction is any fraction whose numerator and/or denominator is also a fraction.

3. Multiplying any expression by _____ does not change its value.

Practice Problems

Objective A Simplify complex fractions whose numerators and/or denominators are rational numbers.

Simplify the following complex fractions.

1. $\dfrac{\frac{1}{7}}{\frac{3}{14}}$

1._____

2. $\dfrac{\frac{9}{8}}{\frac{15}{16}}$

2._____

Name:
Instructor:

Date:
Section:

3. $\dfrac{\dfrac{17}{18}}{\dfrac{51}{72}}$

3. _____

4. $\dfrac{\dfrac{7}{9}}{24}$

4. _____

5. $\dfrac{6}{\dfrac{3}{4}}$

5. _____

6. $\dfrac{\dfrac{1}{2}-\dfrac{1}{3}}{\dfrac{2}{5}}$

6. _____

7. $\dfrac{\dfrac{3}{4}+\dfrac{2}{3}}{\dfrac{1}{6}-\dfrac{1}{9}}$

7. _____

8. $\dfrac{5-\dfrac{3}{7}}{\dfrac{1}{14}+\dfrac{20}{21}}$

8. _____

9. $\dfrac{\dfrac{8}{11}+\dfrac{1}{2}}{2-\dfrac{1}{4}}$

9._____

10. $\dfrac{4-\dfrac{4}{9}}{5-\dfrac{7}{9}}$

10._____

11. $\dfrac{\dfrac{2}{7}+\dfrac{4}{9}}{\dfrac{27}{28}+\dfrac{2}{3}}$

11._____

Objective B Simplify complex fractions whose numerators and/or denominators are rational expressions.

Simplify the following complex fractions.

12. $\dfrac{\dfrac{x}{8}}{\dfrac{y}{6}}$

12._____

13. $\dfrac{\dfrac{a}{20}}{\dfrac{b}{30}}$

13._____

14. $\dfrac{p}{\dfrac{x}{y}}$

14._____

Name: Date:
Instructor: Section:

15. $\dfrac{\dfrac{a}{b}}{c}$ 15._____

16. $\dfrac{\dfrac{a}{b^4}}{\dfrac{c^2}{b^5}}$ 16._____

17. $\dfrac{\dfrac{x^3 y^4}{z^9}}{\dfrac{x^2 y^6}{z^7}}$ 17._____

18. $\dfrac{2+\dfrac{x}{5}}{3-\dfrac{x}{5}}$ 18._____

19. $\dfrac{a-\dfrac{a}{4}}{6+\dfrac{a}{4}}$ 19._____

20. $\dfrac{\dfrac{2}{b}-3}{\dfrac{4}{b^2}-9}$ 20._____

Copyright © 2013 Pearson Education, Inc.

21. $\dfrac{1 + \dfrac{1}{p} - \dfrac{6}{p^2}}{1 - \dfrac{7}{p} + \dfrac{10}{p^2}}$

21._____

22. $\dfrac{\dfrac{6}{x+3} - x}{\dfrac{12}{x+3} + 2}$

22._____

23. $\dfrac{\dfrac{25}{y^2 - 25}}{\dfrac{8}{y+5} - \dfrac{8}{y-5}}$

23._____

24. $\dfrac{\dfrac{5m}{m+3} + \dfrac{2}{m}}{7m - \dfrac{1}{m+3}}$

24._____

25. $\dfrac{\dfrac{q^2 + 10q + 16}{q^2 + 3q}}{\dfrac{q^2 + 11q + 18}{q+3}}$

25._____

Name:
Instructor:
Date:
Section:

Chapter 7 ADDITION AND SUBTRACTION OF RATIONAL NUMBERS AND EXPRESSIONS

7.6 Solving Equations Containing Rational Numbers and Expressions

Learning Objectives
A Solve equations that contain rational numbers.
B Solve equations that contain rational expressions.

Getting Ready

Find the LCD for each of the following:

1. $\dfrac{3}{4}, \dfrac{1}{8}, \dfrac{11}{12}$ 1._____

2. $\dfrac{5}{3x+3}, \dfrac{7}{4x+4}, \dfrac{11}{12}$ 2._____

3. $\dfrac{8}{x^2+9x+20}, \dfrac{10}{x^2+2x-15}, \dfrac{14}{x^2+x-12}$ 3._____

Simplify the following, using the distributive property:

4. $18\left(\dfrac{4x-3}{9} - \dfrac{3x+4}{2}\right)$ 4._____

5. $15(x+2)\left(\dfrac{4}{3(x+2)} - \dfrac{5}{5(x+2)}\right)$ 5._____

Solve the following linear equations:

6. $27 - 4x = 5x$ 6._____

7. $8x - 14 = 4x - 22$ 7._____

8. Solve $x^2 - 30 = x$ 8._____

Key Terms

Use the vocabulary terms listed below to complete each statement in exercises 1–3.

LCD   **multiplication property of equality**   **extraneous**

1. The _____ says if $a=b$ and $c \neq 0$ then $ac=bc$.

2. When solving equations that contain fractions or rational expressions, we multiply both sides of the equation by the _____ of the fractions or rational expressions.

3. It is possible to solve an equation containing rational expressions and get a number that does not solve the equation because it makes one or more of the denominators equal to 0. Such a number is called a(n) _____ solution.

Practice Problems

Objective A   Solve equations that contain rational numbers.

Solve the following equations.

1. $\dfrac{x}{4} - \dfrac{1}{3} = \dfrac{7}{12}$ 1._____

2. $\dfrac{y}{2} + \dfrac{5}{6} = 4$ 2._____

3. $\dfrac{a}{3} + \dfrac{5}{9} = \dfrac{17}{18}$ 3._____

4. $\dfrac{1}{8}b - \dfrac{1}{16}b = \dfrac{3}{4}$ 4._____

Name: Date:
Instructor: Section:

5. $\dfrac{2}{7}q + \dfrac{3}{14}q = \dfrac{10}{21}$ 5._____

6. $\dfrac{n}{2} - \dfrac{3}{8} = \dfrac{5n}{4} + \dfrac{7}{2}$ 6._____

7. $6x - 5 = \dfrac{7x+10}{3}$ 7._____

8. $\dfrac{8x+3}{32} - \dfrac{3x+5}{16} = -\dfrac{7}{32}$ 8._____

9. $\dfrac{4y+1}{24} - \dfrac{2y-7}{8} = \dfrac{2}{3}$ 9._____

10. $\dfrac{p+6}{18} + \dfrac{p+4}{6} = \dfrac{16}{9}$ 10._____

Objective B Solve equations that contain rational expressions.

Solve the following equations.

11. $\dfrac{5}{x} + \dfrac{1}{3} = \dfrac{5}{6}$

11._____

12. $\dfrac{2}{7} - \dfrac{3}{m} = \dfrac{5}{14}$

12._____

13. $\dfrac{6}{y} + \dfrac{3}{2y} = 18$

13._____

14. $\dfrac{6}{5q} - \dfrac{3}{4q} = \dfrac{3}{20}$

14._____

15. $\dfrac{8}{x-2} = \dfrac{9}{2} - \dfrac{1}{x-2}$

15._____

Name:
Instructor:

Date:
Section:

16. $\dfrac{d}{d-32} = -\dfrac{1}{d+3}$

16._____

17. $y - 2 = 3 - \dfrac{4}{y}$

17._____

18. $\dfrac{5}{p+4} = 8 - \dfrac{p}{p+2}$

18._____

19. $\dfrac{7}{m+2} - \dfrac{2}{m^2+6m+8} = \dfrac{3}{m+4}$

19._____

20. $\dfrac{x+2}{x-2} + \dfrac{4}{x^2-4} = \dfrac{x+8}{x+2}$

20._____

21. $\dfrac{24}{x^2+5x} - \dfrac{6}{x+5} = 3$

21._____

22. $\dfrac{2}{x^2+5x+6} + \dfrac{x}{x+2} = \dfrac{4}{x+3}$

22._____

23. $\dfrac{2}{x^2+9x+20} + \dfrac{3}{x^2+6x+5} = \dfrac{-1}{x^2+5x+4}$

23._____

24. $\dfrac{w}{w-2} = \dfrac{4w}{5w+7}$

24._____

25. $\dfrac{5}{b+2} - \dfrac{3}{b+5} = \dfrac{9}{b^2+7b+10}$

25._____

Name:
Instructor:

Date:
Section:

Chapter 7 ADDITION AND SUBTRACTION OF RATIONAL NUMBERS AND EXPRESSIONS

7.7 Applications with Rational Expressions

Learning Objectives
A   Solve application problems involving number properties.
B   Solve application problems involving work.
C   Solve application problems involving distance, rate, and time.

Getting Ready

Solve the following equations:

1. $\dfrac{1}{12} + \dfrac{1}{x} = \dfrac{1}{9}$ 1._____

2. $\dfrac{225}{x} = \dfrac{100}{x} - \dfrac{1}{5}$ 2._____

3. $\dfrac{36}{x-6} = \dfrac{18}{x+6}$ 3._____

4. $\dfrac{1}{x} + \dfrac{4}{x+3} = \dfrac{11}{18}$ 4._____

Represent the following, using x as the variable:

5. two consecutive odd integers   5._____

6. three consecutive even integers   6._____

Copyright © 2013 Pearson Education, Inc.   311

Key Terms

Use the vocabulary terms listed below to complete each statement in exercises 1–5.

$x+2 \qquad x \qquad rt \qquad x+1 \qquad \dfrac{1}{x}$

1. If x represents the smallest integer, then _____ represents the next consecutive integer.

2. If x represents the smallest odd integer, then _____ represents the next odd consecutive integer.

3. The reciprocal of the number x, $x \neq 0$ is _____ .

4. If it takes _____ hours to do a task then $\dfrac{1}{x}$ is the part of the task that can be done in 1 hour.

5. Distance $d =$ _____ .

Practice Problems

Objective A Solve application problems involving number properties.

Solve.

1. One integer is 6 more than another integer. If $\dfrac{1}{3}$ of the larger integer is equal to $\dfrac{1}{2}$ of the smaller integer, find the two integers.

   1. _____

2. Two numbers differ by 16. If $\dfrac{3}{4}$ of the smaller number is 2 less than $\dfrac{1}{3}$ of the larger number, find the two numbers.

   2. _____

Name: Date:
Instructor: Section:

3. The sum of the reciprocals of two consecutive odd integers is $\frac{12}{35}$. Find the two consecutive odd integers.

3._____

4. Eleven times the reciprocal of the sum of two consecutive even integers is $\frac{1}{2}$. Find the two consecutive even integers.

4._____

5. A truck costs $\frac{1}{3}$ as much as a van. If together they cost $48,000, how much does the van cost?

5._____

6. The numerator and the denominator of a fraction are the same. If 12 is subtracted from the denominator, the new fraction is equivalent to $\frac{13}{7}$. Find the original numerator.

6._____

7. One number is four times another. The sum of their reciprocals is $\frac{1}{4}$. Find the numbers.

7._____

8. Two divided by the sum of x and 3 equals the quotient of 3 and the difference of 5 and x. Find x.

8._____

Objective B Solve application problems involving work.

Solve.

9. Scott can wash the fleet of company cars in 8 hours. Marisol can wash the same fleet of cars in 6 hours. Working together, how fast can they wash the fleet of company cars?

9._____

10. Wanda can clean a suite of offices in 6 hours, and her sister Alice can do the same cleaning job in 9 hours. How long would it take if they worked together?

10._____

11. Working together, Joanne and Baxter can prepare refreshments for a youth camp in 40 minutes. If it takes Joanne 70 minutes to prepare the refreshments alone, how long would it take Baxter to prepare the refreshments alone?

11._____

Name: Date:
Instructor: Section:

12. An experienced bricklayer constructs a walkway in 5 hours. His apprentice completes the same type of job in 8 hours. Find how long it takes if they work together.

12._____

13. A pool is being filled by two pipes. Using both pipes it takes 20 hours to fill the pool, while using pipe A alone it takes 30 hours. How long would it take pipe B alone to fill the pool?

13._____

14. Jenna takes 7 minutes longer to print a batch of brochures than Willy, and together they can do the job in 9 minutes less than Willy alone. How long does it take Jenna and Willy working together to print the batch of brochures?

14._____

Objective C   Solve application problems involving distance, rate, and time.

Solve.

15. It took Martin 7 hours to drive 410 miles. Before noon he was averaging 65 miles per hour while after noon it started to rain and he only averaged 50 miles per hour. How much time did he drive after noon?

15._____

16. Cory can row 6 miles downstream in the same amount of time he rows 4 miles upstream. If the speed of the river's current is 3 miles per hour, find the speed of the boat in still water.

16._____

17. Camela and Sarah decide to drive to Virginia. They both leave from the same place at the same time, but Sarah drives 5 miles per hour slower than Camela. If Camela has traveled 65 miles in the same time Sarah has traveled 60 miles, find how fast each was driving.

17._____

18. A truck makes a trip of 420 miles in 4 hours less time than it takes the same van to make a trip of 660 miles traveling at the same speed. What is the speed of the van?

18._____

19. A plane flies against the wind from Florida to Maryland in $3\frac{1}{2}$ hours. It flies back to Florida with the wind in 3 hours. If the average speed of the wind is 31 miles per hour, what is the speed of the plane in still air?

19._____

20. Consuela and Abby start jogging around the track together at the same time. Abby jogs 2 miles per hour faster than Consuela. If Abby can jog 6 miles in the same amount of time it takes Consuela to jog 4 miles, find how fast Abby jogs.

20._____

Name: Date:
Instructor: Section:

Chapter 8   RATIOS, PERCENTS, AND APPLICATIONS

8.1   Ratios and Rates

Learning Objectives
A   Recognize the ratio of two or more numbers.
B   Simplify ratios to lowest terms.
C   Write ratios so that each part is a natural number.
D   Recognize the rate of two measurements.
E   Recognize and use unit pricing.

Getting Ready

1. Write 72 as a product of prime factors.   1._____

2. Reduce $\frac{18}{54}$ to lowest terms.   2._____

3. Multiply $\frac{2}{9} \cdot \frac{4}{9}$ 3._____

4. Simplify $\frac{\frac{1}{9}}{\frac{4}{9}}$.   4._____

Key Terms

Use the vocabulary terms listed below to complete each statement in exercises 1–4.

     rates   ratio   unit price   common factors

1. Ratios are in lowest terms when the parts of a ratio are integers that contain no_____ other than the number 1.

2. Ratios that have different units of measurement are called _____.

3. A(n)_____ is a comparison of two or more quantities by division.

4. A rate that expresses the cost compared to one unit is called a(n) _____.

Practice Problems

Objective B Simplify ratios to lowest terms.

Simplify the following ratios.

1. 9 to 36

   1._____

2. 16 to 26

   2._____

3. 15:95

   3._____

4. 14:18

   4._____

5. $\dfrac{80}{114}$

   5._____

6. $\dfrac{122}{60}$

   6._____

Write each of the following as a ratio in simplest form with each part of the ratio as a natural number.

7. A math class has 18 females and 16 males. What is the ratio of females to males?

   7._____

8. A parking lot has 20 motorcycles and 110 cars. What is the ratio of motorcycles to cars?

   8._____

9. Lemonade is 2 parts lemon juice to 12 parts water. What is the ratio of lemon juice to water?

   9._____

10. Wendy looked at her transcript and noticed she had 12 As and 10 Bs. What is her ratio of As to Bs?

    10._____

Name:
Instructor:

Date:
Section:

Objective C Write ratios so that each part is a natural number.

Write each of the following as a ratio in simplest form with each part of the ratio as a natural number.

11. A crepe recipe calls for $1\frac{1}{2}$ cups of flour to 2 cups of milk. What is the ratio of flour to milk?

    11._____

12. Janet pays .08 of a dollar in tax for every dollar she spends on books at a bookstore. What is the ratio of tax dollars to dollars she spends on books?

    12._____

13. Amy has a special recipe for a fruit salad. Peaches make up $\frac{1}{6}$ of the recipe while pears make up $\frac{3}{16}$ of the recipe. What is the ratio of pears to peaches?

    13._____

14. Seniors make up $\frac{3}{10}$ of the students at a movie while juniors make up $\frac{5}{11}$ of the crowd. What is the ratio of juniors to seniors?

    14._____

Objective D Recognize the rate of two measurements.

Express each rate with the second part equal to 1.

15. Trung drove 495 miles in 9 hours. At what rate did he drive his car?

    15._____

16. If 15 students must share 90 resource books, how many resource books are there per student?

    16._____

17. A family of four people won $125,000 in a contest. What were the average winnings per person?

    17._____

Objective E Recognize and use unit pricing.

Express each of the following as a unit price. If necessary, round answers to the nearest cent.

18. Eight apples cost $1.60. What is the cost per apple?

    18._____

19. A bottle of 250 vitamins cost $4.50. What is the cost per tablet?

    19._____

20. A box containing 32 ounces of cereal costs $4.95. What is the cost per ounce?

    20._____

21. Fifteen tickets to a show cost $41.25. What is the cost per ticket?

    21._____

Determine the unit price (to the nearest tenth of a cent) for each of the different sizes and indicate how much less the larger size is per unit.

22. A 14-ounce bag of vegetables sells for $1.79, and an 18-ounce bag sells for $2.05.

    22._____

23. A 16-ounce box of cookies sells for $2.98, and a 24-ounce box sells for $3.60.

    23._____

24. An $8\frac{1}{2}$-ounce can of peaches sells for $1.10, and a 15-ounce can sells for $1.55.

    24._____

Name:
Instructor:

Date:
Section:

Chapter 8　RATIOS, PERCENTS, AND APPLICATIONS

8.2　Proportions

Learning Objectives
A　Verify proportions.
B　Solve proportions.
C　Solve application problems by using proportions.

Getting Ready

1. Find the product: 7·54

1._____

Solve the following linear equations:

2. $5x = 90$

2._____

3. $5x + 4 = 8x + 1$

3._____

4. Solve the quadratic equation $x^2 - x - 9 = 21$.

4._____

5. Find the product: 6.1(3)

5._____

6. Find the quotient: $\dfrac{112}{3.5}$

6._____

7. Find the product: $16 \cdot \dfrac{9}{4}$

7._____

8. Solve $d = rt$ for t.

8._____

Key Terms

Use the vocabulary terms listed below to complete each statement in exercises 1–2.

proportion　　　　　**cross products**

1. A proportion is true if its _____ are equal.

2. A _____ is an equation that states that two or more ratios are equal.

Copyright © 2013 Pearson Education, Inc.　　321

Practice Problems

Objective A Verify proportions.

Determine whether the following fractions can form a proportion.

1. $\dfrac{5}{9}$ and $\dfrac{20}{36}$

    1._____

2. $\dfrac{2}{7}$ and $\dfrac{18}{40}$

    2._____

3. $\dfrac{11}{3}$ and $\dfrac{64}{18}$

    3._____

4. $\dfrac{6}{40}$ and $\dfrac{15}{100}$

    4._____

Objective B Solve proportions.

Solve the following proportions.

5. $\dfrac{6}{7} = \dfrac{12}{x}$

    5._____

6. $\dfrac{y}{9} = \dfrac{8}{36}$

    6._____

7. $\dfrac{110}{120} = \dfrac{a}{660}$

    7._____

8. $\dfrac{14}{b} = \dfrac{63}{9}$

    8._____

Name:
Instructor:

Date:
Section:

9. $\dfrac{5}{7} = \dfrac{25}{x}$

9. _____

10. $\dfrac{3}{4} = \dfrac{36}{y}$

10. _____

11. $\dfrac{\frac{9}{7}}{8} = \dfrac{m}{112}$

11. _____

12. $\dfrac{a}{60} = \dfrac{.4}{12}$

12. _____

13. $\dfrac{3.6}{p} = \dfrac{9}{10}$

13. _____

14. $\dfrac{2\frac{1}{8}}{6} = \dfrac{q}{96}$

14. _____

15. $\dfrac{x+4}{6} = \dfrac{7}{3}$

15. _____

16. $\dfrac{x-1}{32} = \dfrac{1}{2}$

16. _____

17. $\dfrac{-x}{6} = \dfrac{1}{x-5}$

17. _____

18. $\dfrac{4}{3x-7} = \dfrac{-6}{5x+1}$

18. _____

19. $\dfrac{8}{4x-4} = \dfrac{10}{2x+1}$

19. _____

20. $\dfrac{2x+1}{-3} = \dfrac{1}{3x-4}$

20._____

Objective C   Solve application problems by using proportions.

Solve each of the following.

21. A car travels 360 miles in 6 hours. How far will the car travel in 9 hours?

21._____

22. Milly can type 304 words in 4 minutes. How many words can she type in 11 minutes?

22._____

23. On a blueprint 1 inch corresponds to 10 feet. Find the length of a wall represented by a line $3\dfrac{2}{3}$ inches long on the blueprint.

23._____

24. Paul's car can go 352 miles on 16 gallons of gas. How much gas can he expect to use on a 682 mile trip?

24._____

25. Three bags of grass seed cover 3600 square feet of lawn. How many bags are required to cover 4500 square feet of lawn?

25._____

Name:
Instructor:

Date:
Section:

Chapter 8 RATIOS, PERCENTS, AND APPLICATIONS

8.3 Percent

Learning Objectives
A   Convert percent notation to fractions.
B   Convert percent notation to decimals.
C   Convert decimals to percent notation.
D   Convert fractions to percent notation.
E   Add and subtract in percent notation.

Getting Ready

1. Write $\dfrac{14}{100,000}$ as a decimal. 1._____

2. Change $3\dfrac{6}{11}$ into an improper fraction. 2._____

Find the following products:

3. $\dfrac{19}{5} \cdot \dfrac{1}{100}$ 3._____

4. $11.8(.01)$ 4._____

5. Divide $\dfrac{660}{9}$ and round the answer to the nearest hundredth. 5._____

Add or subtract the following:

6. $7.6 - 5$ 6._____

7. $4 + 7\dfrac{1}{2}$ 7._____

Copyright © 2013 Pearson Education, Inc.

Key Terms

Use the vocabulary terms listed below to complete each statement in exercises 1–3.

.01 $\dfrac{1}{100}$ **percent**

1. _____ notation is a special way of writing a fraction with a denominator of 100.

2. To convert from percent notation to a decimal replace 1% with_____ and multiply.

3. To convert from percent notation to a fraction replace 1% with_____ and multiply.

Practice Problems

Objective A Convert percent notation to fractions.

Convert each of the following percents to fractions.

1. 18%   1._____

2. 6%   2._____

3. 170%   3._____

4. 28.9%   4._____

5. $3\dfrac{1}{4}\%$ 5._____

6. $13\dfrac{7}{9}\%$ 6._____

Name:
Instructor:

Date:
Section:

Objective B Convert percent notation to decimals.

Convert the given percents to decimals. If necessary, round to the nearest thousandths.

7. 81%

7._____

8. 42%

8._____

9. .7%

9._____

10. 214%

10._____

11. 241%

11._____

12. 39.4%

12._____

13. 5.06%

13._____

14. $8\frac{3}{4}\%$

14._____

15. $10\frac{5}{12}\%$

15._____

Objective C Convert decimals to percent notation.

Convert the following decimals to percent notation.

16. .8

17. .47

18. .29

19. .0007

20. .041

21. 4.98

22. 12

23. .444…

24. 6

16._____

17._____

18._____

19._____

20._____

21._____

22._____

23._____

24._____

Name: Date:
Instructor: Section:

Objective D Convert fractions to percent notation.

Convert the following fractions to percent notation using the method that works best for each fraction.

25. $\dfrac{1}{5}$ 25._____

26. $\dfrac{9}{10}$ 26._____

27. $\dfrac{3}{8}$ 27._____

28. $\dfrac{13}{50}$ 28._____

29. $\dfrac{8}{7}$ 29._____

30. $\dfrac{5}{12}$ 30._____

Objective E Add and subtract in percent notation.

Perform each indicated operation.

31. $13\% + 26\%$

31._____

32. $49\% - 17\%$

32._____

33. $51.7\% + 19\%$

33._____

34. $4\frac{3}{8}\% - 2\frac{1}{5}\%$

34._____

Name:
Instructor:

Date:
Section:

Chapter 8 RATIOS, PERCENTS, AND APPLICATIONS

8.4 Applications of Percent

Learning Objectives
A   Translate mathematical expressions involving percent into English.
B   Solve simple percent problems.
C   Solve simple percent problems with proportions.
D   Solve application problems involving percent.

Getting Ready

1. Convert 70% into a fraction.   1._____

2. Convert 8% into a decimal.   2._____

Multiply or divide the following decimals:

3. $.232(18,000)$ 3._____

4. $\dfrac{180}{.04}$ 4._____

Solve the following equations:

5. $.8x = 64$ 5._____

6. $5.2y = 338$ 6._____

Solve the following proportions:

7. $\dfrac{2}{3} = \dfrac{P}{162}$ 7._____

8. $\dfrac{A}{100} = \dfrac{9}{5}$ 8._____

Copyright © 2013 Pearson Education, Inc.   331

Key Terms

Use the vocabulary terms listed below to complete each statement in exercises 1–5.

| base | number of percent | amount | equals | multiplication |

1. In the proportion $\dfrac{x}{100} = \dfrac{y}{z}$, y represents the _____, or part.

2. The word "is" translates as _____.

3. In the proportion $\dfrac{x}{100} = \dfrac{y}{z}$, z represents the _____.

4. The word "of" translates as _____.

5. In the proportion $\dfrac{x}{100} = \dfrac{y}{z}$, x represents the _____.

Practice Problems

Objective A Translate mathematical expressions involving percent into English.

Translate the following percent equations into English sentences.

1. $x = 14\% \cdot 38$ 1._____

2. $35\% \cdot 22 = x$ 2._____

3. $22\% \cdot y = 7$ 3._____

4. $t\% \cdot 120 = 18$ 4._____

Objective B Solve simple percent problems.

Solve the following by using simple percent sentences.

5. 8% of 600 is what number?   5._____

Name: Date:
Instructor: Section:

6. What number is 12% of 70? 6._____

7. 14 is 80% of what number? 7._____

8. 24.65 is 42.5% of what number? 8._____

9. 14 is what percent of 3.5? 9._____

10. 12 is what percent of 300? 10._____

11. $15\frac{1}{3}$% of what number is 92? 11._____

12. $20\frac{3}{4}$% of 190 is what number? 12._____

Objective C   Solve simple percent problems with proportions.

Solve the following by using proportions.

13. What number is 15% of 90? 13._____

14. 82% of 350 is what number? 14._____

15. 32.5 is what percent of 162.5? 15._____

16. 112 is what percent of 80? 16._____

17. What number is 30% of 160?   17._____

18. 50% of what number is 72?   18._____

19. $15\frac{2}{3}\%$ of 180 is what number?   19._____

20. $2\frac{1}{8}\%$ of what number is 34?   20._____

Objective D Solve application problems involving percent.

Solve each of the following and round the answers to the nearest percent or nearest hundredth where appropriate.

21. Randy has completed 65 pages of his math workbook that has 195 pages. What percent of the workbook has he finished?   21._____

22. A mixture of fertilizer and soil contains 8% fertilizer. How many pounds of fertilizer are in 55 pounds of this mixture?   22._____

23. By volume a certain drink is 16% pure juice. If a bottle contains 64 ounces of this drink, how many ounces of pure juice are in the bottle?   23._____

24. Phyllis buys 6 pounds of hamburger that claims to be 96% lean (fat free). How many pounds of fat are in her purchase?   24._____

Name: Date:
Instructor: Section:

Chapter 8 RATIOS, PERCENTS, AND APPLICATIONS

8.5 Further Applications of Percent

Learning Objectives
A Compute sales tax.
B Compute discount.
C Compute commission.
D Compute simple interest.
E Compute compound interest.
F Calculate percent increase and percent decrease.

Getting Ready

1. Solve $.08P = 16$ for P.

   1._____

2. Change 72% into a decimal.

   2._____

3. Change 24% into a fraction.

   3._____

4. Multiply: $.035(220)$

   4._____

5. Divide: $\dfrac{18}{.04}$

   5._____

6. Solve: $\dfrac{14}{100} = \dfrac{98}{p}$

   6._____

7. Multiply: $1460(.18)\left(\dfrac{1}{2}\right)$

   7._____

8. Change $\dfrac{3}{5}$ into a percent.

   8._____

Key Terms

Use the vocabulary terms listed below to complete each statement in exercises 1–8.

commission	purchase price	sales	sales tax
total cost	principal	discount	compound

1. _____ interest is interest computed on previously earned interest in addition to principal.

2. _____ is the amount of money that we pay to the government when we make certain types of purchases.

3. A _____ is the amount of money that is paid to a person for selling an item.

4. Sales tax rate · _____ = sales tax.

5. A _____ is the amount of money that is taken off the original price of an item.

6. _____ = purchase amount + sales tax.

7. The amount of money borrowed or deposited at a bank is called the _____.

8. Commission rate · _____ = amount of commission.

Practice Problems

Objective A Compute sales tax.

Solve the following involving sales tax.

1. Find the tax on an item that sells for $45 if the tax rate is 6%.

   1._____

2. If the tax on a television that sells for $1240 is $148.80, find the tax rate.

   2._____

3. The sales tax on a stereo is $27.20. If the sales tax rate is 8%, what is the purchase amount?

   3._____

Name: Date:
Instructor: Section:

4. The sales tax rate in a particular state is 4.5%. If you purchase a bedroom dresser for $640, what is the total cost?

4._____

Objective B Compute discount.

Solve the following involving discounts.

5. If an item has an original price of $160 and the discount rate is 15%, find the discount.

5._____

6. A skirt has an original price of $62 and is discounted by $18.60. What is the rate of discount?

6._____

7. Donna paid $980 for a rug that was discounted $200. What was the original price of the rug? What was the discount rate?

7._____

Objective C Compute commission.

Solve the following involving commission.

8. If the commission rate is 6%, find the commission on sales of $550.

8._____

9. Melissa received $42 commission for selling a chair that sold for $350. What was the commission rate?

9._____

10. Gary works for 20% commission. If he received a commission of $96 on the sale of furniture, what was the selling price?

10._____

Objective D Compute simple interest.

Given the interest formula $I = p \cdot r \cdot t$, solve each exercise for the unknown value. If necessary, round to the nearest cent, percent, or year.

11. $r = 8\%, p = \$300, t = 4$ years, $I =$

11. _____

12. $r = 5.5\%, I = \$330, t = \frac{2}{3}$ year, $p =$

12. _____

13. $r = 6\%, I = \$550.80, p = \$1020, t =$

13. _____

14. $p = \$2400, I = \$588, t = 7$ years, $r =$

14. _____

Calculate the simple interest to the nearest cent for each of the following sets of information.

15. $p = \$7000, r = 4.5\%, t = 30$ days

15. _____

16. $p = \$5200, r = 15.75\%, t = 8$ months

16. _____

17. $p = \$1250, r = 10\frac{1}{2}\%, t = 10$ months

17. _____

Name:
Instructor:

Date:
Section:

Objective E Compute compound interest.

Calculate the interest to the nearest cent.

18. Calculate the balance after 1 year for an account that pays 6% compounded semiannually on a beginning balance of $4000.

    18._____

19. Calculate the balance after 1 year for an account that pays 7% compounded semiannually on a beginning balance of $1200.

    19._____

Objective F Calculate percent increase and percent decrease.

Solve the exercises using percent increase and percent decrease. If necessary, round to the nearest tenth.

20. Pam's puppy weighed 16 pounds at his previous vet check. At today's check the puppy weighed 22 pounds. What is the percent of increase in his weight?

    20._____

21. The average house value in Augusta County went from $180,000 last year to $225,000 this year. What is the percent of increase in house value?

    21._____

22. Vito's Pizzeria experienced a 3% decrease in sales. If the total sales after the decrease were $28,000, what were the sales before the decrease?

    22._____

23. A cold front went through a town causing the temperature to drop 32%. If the temperature was 80° before the front, what was it after the temperature fell?

23._____

24. Furniture Barn currently produces 8400 chairs per month. If production decreases 9%, find the new number of chairs produced per month.

24._____

25. Juan decided to decrease the number of calories in his diet from 2800 to 2200. Find the percent of decrease.

25._____

Name:
Instructor:

Date:
Section:

Chapter 9   SYSTEMS OF LINEAR EQUATIONS

9.1   Defining Linear Systems and Solving by Graphing

Learning Objectives
A   Determine whether a given ordered pair is the solution of a linear system with two variables.
B   Solve a linear system with two variables by graphing.
C   Determine whether a linear system with two variables is consistent, inconsistent, or dependent.

Getting Ready

Determine whether the given ordered pair is a solution to the given equation.

1. $2x - y = 5$; $(3,1)$

1._____

2. $4x + 3x = 12$; $(0,-4)$

2._____

Given a value of x or y, determine the other coordinate that will make the ordered pair lie on the line represented by the given equation.

3. $x - 5y = 9$; $y = 4$

3._____

4. $3x + 4y = 8$; $x = 4$

4._____

Are the following pairs of line parallel?

5. $3x - 5y = 6$
    $3x + 5y = -6$

5._____

6. $x - 6y = 7$
    $x - 6y = 8$

6._____

Copyright © 2013 Pearson Education, Inc.   341

Key Terms

Use the vocabulary terms listed below to complete each statement in exercises 1–6.

coordinates **inconsistent** **system** **consistent** **dependent** **solution**

1. If a linear system with two variables has one solution the system is said to be_____.

2. If a linear system with two variables has no solution the system is said to be_____.

3. If a linear system with two variables has an infinite number of solutions the system is said to be_____.

4. Any ordered pair that solves all the equations in a system of equations is a(n) _____ of the system.

5. A(n) _____ of linear equations is two or more linear equations with the same variables.

6. To solve a system of linear equations with two variables, graph each equation and if the graphs intersect, the _____ of the point(s) of intersection are the solutions of the system.

Practice Problems

Objective A Determine whether a given ordered pair is the solution of a linear system with two variables.

Determine whether the given ordered pair is a solution of the system of linear equations. The ordered pair is in the form (x, y).

1. $(1, 7)$
$$\begin{cases} x + 2y = 15 \\ 2x - y = -5 \end{cases}$$

1._____

2. $(-2, 3)$
$$\begin{cases} 3x + y = -3 \\ -x + y = 1 \end{cases}$$

2._____

Name:
Instructor:

Date:
Section:

3. $(0, 4)$
$$\begin{cases} 5x + 9y = 36 \\ -4x + 2y = -8 \end{cases}$$

3. _____

4. $(6, -5)$
$$\begin{cases} x - 7y = 41 \\ 3x - y = 23 \end{cases}$$

4. _____

5. $(-2, -10)$
$$\begin{cases} x - y = 8 \\ -x - y = 12 \end{cases}$$

5. _____

6. $(10, 15)$
$$\begin{cases} 2x + y = 35 \\ -x + y = -5 \end{cases}$$

6. _____

Objective B Solve a linear system with two variables by graphing.

The graphs shown here are of systems of equations. (a) Give the type of system (consistent, inconsistent, or dependent). (b) Give the number of solutions, if any. (c) If there is only one solution, give the solution.

7. $\begin{cases} x + y = 4 \\ x - y = 2 \end{cases}$

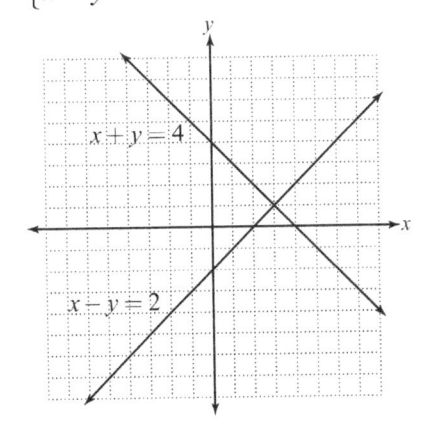

7a. _____

b. _____

c. _____

8. $\begin{cases} 2x+y=0 \\ x+y=2 \end{cases}$

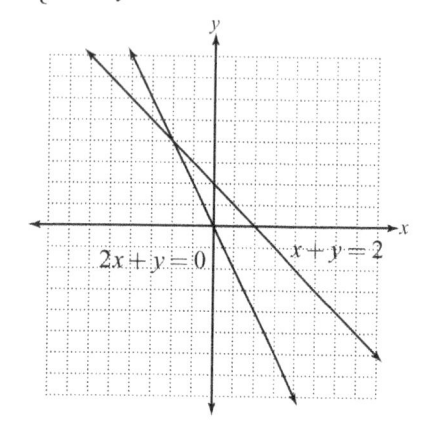

8a._____

b._____

c._____

9. $\begin{cases} x+3y=6 \\ -x-3y=9 \end{cases}$

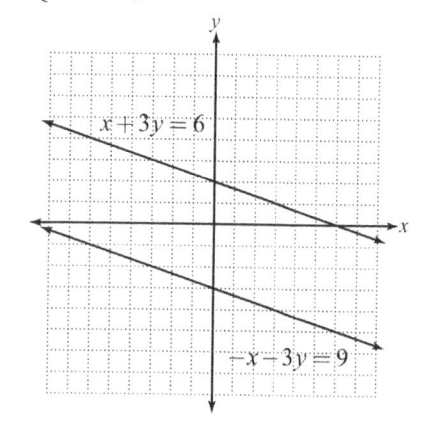

9a._____

b._____

c._____

10. $\begin{cases} 3x+6y=-18 \\ x+2y=-6 \end{cases}$

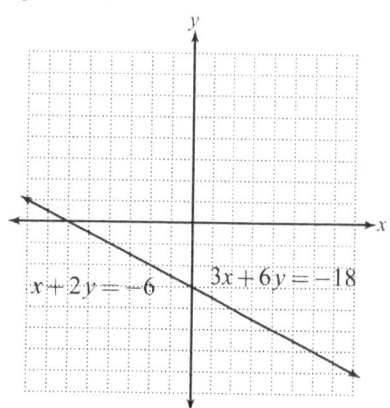

10a._____

b._____

c._____

Name:
Instructor:

Date:
Section:

Find the solution(s) of the given systems of linear equations by graphing. Assume that each unit on the coordinate system represents 1.

11. $\begin{cases} y = 3x + 2 \\ y = -x + 6 \end{cases}$

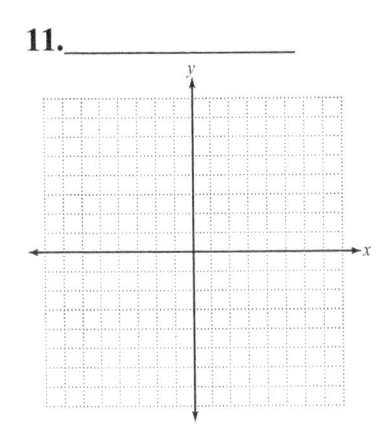

11. _____

12. $\begin{cases} y = \dfrac{1}{2}x - 1 \\ y = -\dfrac{1}{6}x - 5 \end{cases}$

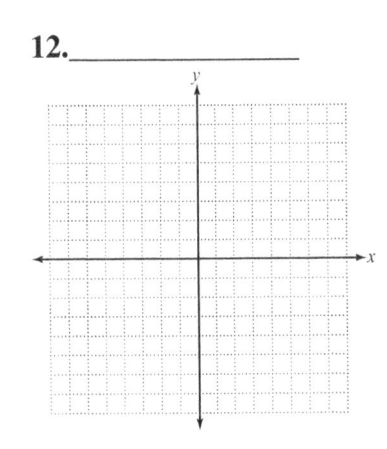

12. _____

13. $\begin{cases} y = 7x + 6 \\ y = 6 \end{cases}$

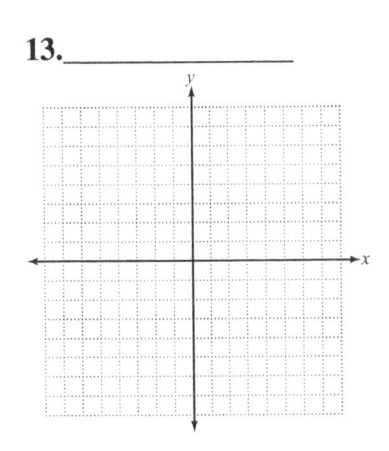

13. _____

14. $\begin{cases} x = 5 \\ y = -7 \end{cases}$

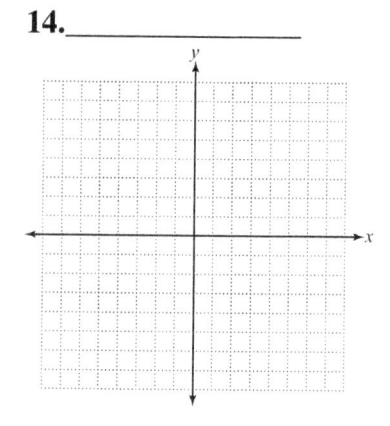

14. _____

15. $\begin{cases} x + y = 7 \\ y - 4 = 0 \end{cases}$

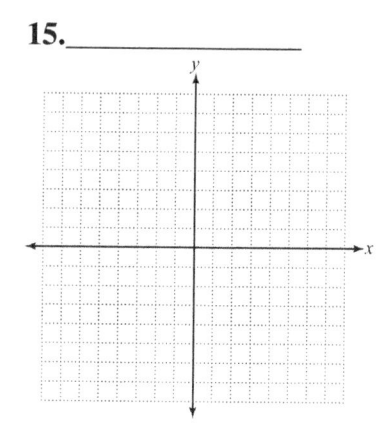

15. _____

16. $\begin{cases} 6x - 4y = 4 \\ 4x + 2y = 12 \end{cases}$

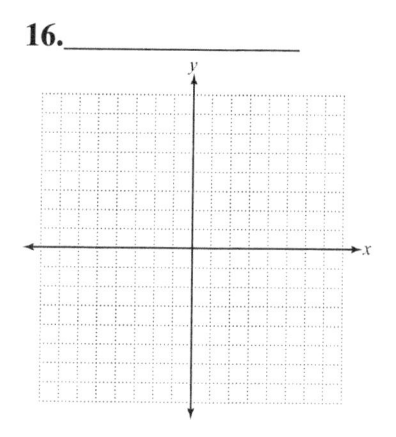

16. _____

Name:
Instructor:

Date:
Section:

17. $\begin{cases} 4x - 5y = 0 \\ 3x + 2y = 0 \end{cases}$

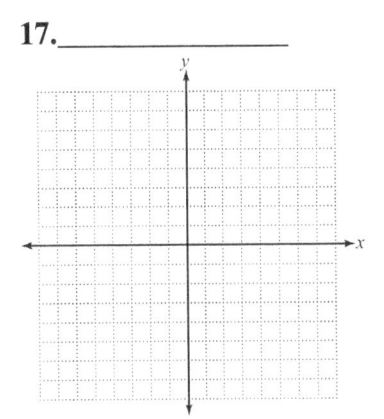

17._____

18. $\begin{cases} x - 4 = 2 \\ y - 5 = 0 \end{cases}$

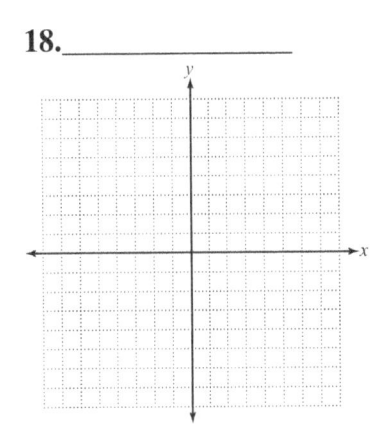

18._____

Objective C Determine whether a linear system with two variables is consistent, inconsistent, or dependent.

Determine whether the systems given are consistent, inconsistent, or dependent. If the system is consistent, find the solution of the system.

19. $\begin{cases} 3x + 4y = 8 \\ 6x + 8y = 16 \end{cases}$

19._____

20. $\begin{cases} 4x - 12y = 8 \\ 10x - 30y = 22 \end{cases}$ 20._____

21. $\begin{cases} -3x + 7y = 2 \\ 6x - 14y = 4 \end{cases}$ 21._____

22. $\begin{cases} 5x + 2y = 3 \\ x - y = 2 \end{cases}$ 22._____

Name:
Instructor:
Date:
Section:

Chapter 9   SYSTEMS OF LINEAR EQUATIONS

9.2   Solving Systems of Linear Equations by Using Elimination by Addition

Learning Objectives
A   Use the addition method to solve consistent systems of linear equations with two variables by eliminating a variable.
B   Recognize inconsistent and dependent systems of equations when using the addition method.

Getting Ready

Add, subtract, multiply, or divide as indicated.

1. $10+(-8)$

   1._____

2. $9-(-7)$

   2._____

3. $(-3)(9)$

   3._____

4. $12 \div (-4)$

   4._____

Combine like terms.

5. $6x+7-4x$

   5._____

6. $-3a-1+2a+5$

   6._____

7. Is $(2,-5)$ a solution of $x+y=-3$?

   7._____

8. Multiply: $\left(-\dfrac{1}{4}\right)(16)$

   8._____

9. Rewrite $y=-6x+4$ in the $ax+by=c$ form.

   9._____

10. If $y=-\dfrac{1}{5}$ and $x+10y=-9$, find x.

    10._____

Copyright © 2013 Pearson Education, Inc.

11. Simplify $7(5x-y)$ using the distributive property.   11._____

Key Terms

Use the vocabulary terms listed below to complete each statement in exercises 1–3.

    additive inverses   **Addition of Equals**   LCD

1. The _____ rule states that if $a=b$ and $c=d$, then $a+c=b+d$.

2. Two numbers are _____ if their sum is 0.

3. When an equation has fractions they may be cleared by multiplying both sides of the equation by the _____ of the fractions.

Practice Problems

Objective A   Use the addition method to solve consistent systems of linear equations with two variables by eliminating a variable.

Solve the following systems of equations by using elimination by addition.

1. $\begin{cases} x+y=6 \\ x-y=4 \end{cases}$ 1._____

2. $\begin{cases} -x+3y=8 \\ x+2y=12 \end{cases}$ 2._____

Name:
Instructor:

Date:
Section:

3. $\begin{cases} 3x+y=3 \\ -x+2y=6 \end{cases}$

3._____

4. $\begin{cases} 5x-y=-2 \\ 4x+3y=-13 \end{cases}$

4._____

5. $\begin{cases} 5x+4y=11 \\ -x+3y=-25 \end{cases}$

5._____

6. $\begin{cases} 2x+7y=6 \\ 5x-3y=-26 \end{cases}$

6._____

7. $\begin{cases} -6x+5y=-48 \\ 4x+2y=32 \end{cases}$ 7._____

8. $\begin{cases} 9x+10y=-97 \\ -8x-7y=73 \end{cases}$ 8._____

9. $\begin{cases} 5x+4y=43 \\ 11x+6y=61 \end{cases}$ 9._____

10. $\begin{cases} 14x+9y=24 \\ 7x-12y=45 \end{cases}$ 10._____

Name:
Instructor:

Date:
Section:

11. $\begin{cases} 4x - 5y = -28 \\ 6x + 7y = 45 \end{cases}$

11._____

12. $\begin{cases} 2x + 3y = 3 \\ 5x - 6y = -15 \end{cases}$

12._____

13. $\begin{cases} 10x + 6y = 7 \\ -5x - 12y = -2 \end{cases}$

13._____

14. $\begin{cases} .3x + .5y = -1.4 \\ 2.6x - 1.3y = -6.5 \end{cases}$

14._____

15. $\begin{cases} \dfrac{3}{4}x - \dfrac{11}{8}y = -11 \\ \dfrac{1}{3}x + \dfrac{7}{4}y = 14 \end{cases}$

15._____

16. $\begin{cases} \dfrac{3}{2}x - \dfrac{1}{3}y = -\dfrac{55}{6} \\ \dfrac{9}{14}x + \dfrac{3}{8}y = -6 \end{cases}$

16._____

Objective B Recognize inconsistent and dependent systems of equations when using the addition method.

Solve the given systems of equations. Indicate whether the system is consistent, inconsistent, or dependent.

17. $\begin{cases} 6x - 5y = 8 \\ -12x + 10y = -16 \end{cases}$

17._____

Name:
Instructor:

Date:
Section:

18. $\begin{cases} 4x+10y=8 \\ 6x+15y=14 \end{cases}$

18._____

19. $\begin{cases} 4y=9x+18 \\ 45x-20y=90 \end{cases}$

19._____

20. $\begin{cases} x=7y-8 \\ 4x-28y=-32 \end{cases}$

20._____

21. $\begin{cases} y=3x+8 \\ -12x-4y=-32 \end{cases}$

21._____

22. $\begin{cases} 5x - y = 14 \\ 10x + 2y = 32 \end{cases}$

22. _____

Name:
Instructor:
Date:
Section:

Chapter 9   SYSTEMS OF LINEAR EQUATIONS

9.3   Solving Systems of Linear Equations by Using Substitution

Learning Objectives
A   Solve systems of linear equations with two variables by using substitution.

Getting Ready

1. Solve $-3x + y = 8$ for y.

    1._____

2. If $x = 5$ and $-2x + y = 12$, find y.

    2._____

3. Is $(-1, 2)$ a solution of $2x + 3y = 6$?

    3._____

Add, subtract, multiply, or divide as indicated.

4. $6 + (-3)$

    4._____

5. $7 - (-8)$

    5._____

6. $(-2)(-6)$

    6._____

7. $24 \div (-4)$

    7._____

8. Multiply: $-7\left(\dfrac{3}{28}\right)$

    8._____

9. Is the system of equations $\begin{cases} -x + 3y = 7 \\ 2x - 6y = -14 \end{cases}$ consistent, dependent, or inconsistent?

    9._____

Copyright © 2013 Pearson Education, Inc.

Key Terms

Use the vocabulary terms listed below to complete each statement in exercises 1–4.

±1   substitution   dependent   inconsistent

1. When using the method of substitution if we arrive at a false statement, we know the system is_____.

2. When using the method of substitution if we arrive at a true statement, we know the system is_____.

3. The method of substitution is most useful when an equation has a variable whose coefficient is_____.

4. The principle of _____ allows us to replace any quantity with any other quantity that is equal to it.

Practice Problems

Objective A   Solve systems of linear equations with two variables by using substitution.

If possible, solve the given systems of equations by the substitution method. If the system is not consistent, indicate whether it is inconsistent or dependent.

1. $\begin{cases} x + 3y = -5 \\ 5x - 9y = 23 \end{cases}$ 1._____

2. $\begin{cases} 8x - y = 28 \\ 4x + 3y = 0 \end{cases}$ 2._____

Name: Date:
Instructor: Section:

3. $\begin{cases} x+y=-11 \\ 2x-y=-4 \end{cases}$ 3._____

4. $\begin{cases} 6x-y=34 \\ 3x+4y=53 \end{cases}$ 4._____

5. $\begin{cases} 7x+3y=42 \\ x+5y=70 \end{cases}$ 5._____

6. $\begin{cases} 8x+9y=-1 \\ -x+3y=-37 \end{cases}$ 6._____

7. $\begin{cases} -2x+7y=9 \\ 10x-y=-79 \end{cases}$ 7._____

8. $\begin{cases} 5x+8y=39 \\ -x+7y=18 \end{cases}$ 8._____

9. $\begin{cases} 15x+9y=-12 \\ 7x-y=-16 \end{cases}$ 9._____

10. $\begin{cases} 6x-y=89 \\ 5x+y=87 \end{cases}$ 10._____

Name: Date:
Instructor: Section:

11. $\begin{cases} -9x + 7y = 21 \\ x + 12y = -79 \end{cases}$

11._____

12. $\begin{cases} 3x - 14y = -18 \\ 6x + y = 51 \end{cases}$

12._____

13. $\begin{cases} 6x + y = 5 \\ -4x + 5y = 8 \end{cases}$

13._____

14. $\begin{cases} 6x + 8y = 10 \\ -3x + 12y = 7 \end{cases}$

14._____

15. $\begin{cases} x - 9y = 18 \\ -3x + 27y = -54 \end{cases}$

15._____

16. $\begin{cases} y = 5x + 17 \\ 10x - 2y = 34 \end{cases}$

16._____

17. $\begin{cases} x + 8 = 14y \\ 2x - 28y = 16 \end{cases}$

17._____

18. $\begin{cases} 2x + 19y = -10 \\ 3x - 8y = -15 \end{cases}$

18._____

Name: Date:
Instructor: Section:

19. $\begin{cases} \dfrac{2}{3}x + \dfrac{1}{9}y = \dfrac{1}{6} \\ 12x + 2y = 3 \end{cases}$ **19.** _____

20. $\begin{cases} -2x + 3y = -34 \\ 5x - 7y = 80 \end{cases}$ **20.** _____

Name: Date:
Instructor: Section:

Chapter 9   SYSTEMS OF LINEAR EQUATIONS

9.4   Solving Application Problems by Using Systems of Equations

Learning Objectives
Solve the following types of application problems, using systems of linear equations:
A   Number
B   Money
C   Distance, rate, and time
D   Percent
E   Mixture

Getting Ready

Solve the following systems of equations:

1. $\begin{cases} 2x - 3y = 16 \\ x + 2y = 1 \end{cases}$

   1._____

2. $\begin{cases} x + y = 2 \\ 6x - 2y = -12 \end{cases}$

   2._____

Multiply as indicated.

3. $100(.08)$

   3._____

4. $10(7.2)$

   4._____

Change the following percents to decimals:

5. 5%

   5._____

6. 76%

   6._____

7. 18%

   7._____

Copyright © 2013 Pearson Education, Inc.

Key Terms

Use the vocabulary terms listed below to complete each statement in exercises 1–3.

          variables   **left**   **time**

1. When using systems of linear equations to solve an application you should have the same number of _____ as equations.

2. To change a percent to a decimal, drop the percent sign and move the decimal point two places to the _____.

3. Distance = rate · _____.

Practice Problems

Objective A Solve application problems, using systems of linear equations: Number

Solve.

1. In 2005 the U.S. population that claimed German ancestry was one million more than 16 times the number that claimed Russian ancestry. If together they numbered 52 million people, find the number of people claiming each ancestry. (Source: The World Almanac and Book of Facts 2007)

1._____

2. In 2005, Tiger Woods' professional golf earnings were approximately 238 times the earnings of Ben Hogan in 1946. If twenty times Hogan's earnings are added to Woods' earnings the total is $10,840,000. What were the earnings of each golfer? (Source: The World Almanac and Book of Facts 2007)

2._____

3. Two numbers have a sum of 112 and have a difference of 54. Find the two numbers.

3._____

Name:
Instructor:

Date:
Section:

4. A first number plus four times a second number is 101. Five times the first number less eight times the second number is −83. Find the two numbers.

4._____

Objective B Solve application problems, using systems of linear equations: Money

Solve.

5. Last month Timothy purchased 5 CDs and 8 DVDs for $188. This month he purchased 4 CDs and 3 DVDs for $96. Find the price for each CD and for each DVD.

5._____

6. Christine wanted to landscape her backyard with ferns and hibiscus plants. She paid $16 for each fern and $22 for each hibiscus plant. If she paid $776 for 44 plants, how many of each type did Christine purchase?

6._____

7. Sarah spent $218 on school uniforms for her twins. She bought 10 shirts and 8 pairs of pants. Twice the price of a pair of pants is five dollars more than three times the price of a shirt. How much did each shirt and each pair of pants cost?

7._____

8. Michelle has 480 coins in a jar, all of which are either dimes or quarters. If the total value is $102, how many of each type coin does she have?

8._____

Objective C Solve application problems, using systems of linear equations: Distance, rate, and time

Solve.

9. An airplane can fly 3640 miles with the wind in the same amount of time it flies 2760 miles against the wind. Find the speed of the wind and the speed of the airplane in still air.

9._____

10. Mike was able to row 12 miles downstream on the St. Johns River in $1\frac{1}{2}$ hours. If the return trip upstream took 6 hours, find the rate at which Mike could row in still water, and find the rate of the current.

10._____

11. Betsy and Aaron are in two different cities 690 miles apart. They are going to travel along I-10 until they meet. If Aaron drives 15 mph faster than Betsy and they meet after 6 hours, how fast was Aaron driving?

11._____

12. Jasper drove from his house to Covington at 60 mph and then later returned home along the same route at 50 mph. Find the distance from his house to Covington if his total driving time for the roundtrip was 8.8 hours.

12._____

Name:
Instructor:

Date:
Section:

Objective D Solve application problems, using systems of linear equations: Percent

Solve.

13. Max invested part of his $50,000 inheritance in a mutual fund that pays 8% simple interest per year. The remainder of his money he invested in a certificate of deposit that pays 5.5% simple interest per year. At the end of one year his investments had earned $3400. How much money did he invest in the mutual fund?

13._____

14. Sylvester invested $6800 into two accounts. In one account he ended up with a 12% gain in one year whereas in the other account he had a 5% loss for the year. If overall for the year his earnings were $408, how much did he invest in each account?

14._____

15. How can $120,000 be invested, part at 4% annual simple interest and the remaining part at 8% annual simple interest so that the interest earned by the two accounts will be equal?

15._____

Objective E Solve application problems, using systems of linear equations: Mixture

Solve.

16. How much water should be added to 30 gallons of a solution that is 85% antifreeze in order to get a mixture that is 70% antifreeze?

16._____

17. Tickets for a local football game were $6.75 for students and $10 for adults. A total of 820 tickets were sold for $6217.50. How many student tickets were sold?

17._____

18. The Better Bean has a special blend coffee it creates. Its manager wants to prepare 120 pounds of this blend and wants to sell it for $8.50 per pound. His creation is made by combining a high-quality bean costing $16.50 per pound with a cheaper bean costing $4.50 per pound. How many pounds of each bean will he need to use?

18._____

19. Titus needs to prepare 30 ounces of a 12.4% acid solution. To get this solution, find the amount of 6% and the amount of 18% solution he should mix.

19._____

20. This year's annual fall festival dinner at Grace Church had 260 people in attendance. They charged $2.25 for children and $4.75 for adults. If they took in $1047.50, find how many adults attended the dinner.

20._____

Name:
Instructor:

Date:
Section:

Chapter 9   SYSTEMS OF LINEAR EQUATIONS

9.5   Systems of Linear Inequalities

Learning Objectives
A   Graphically represent the solutions of a system of linear inequalities.

Getting Ready

Graph the following inequalities:

1. $x < -7$

1.

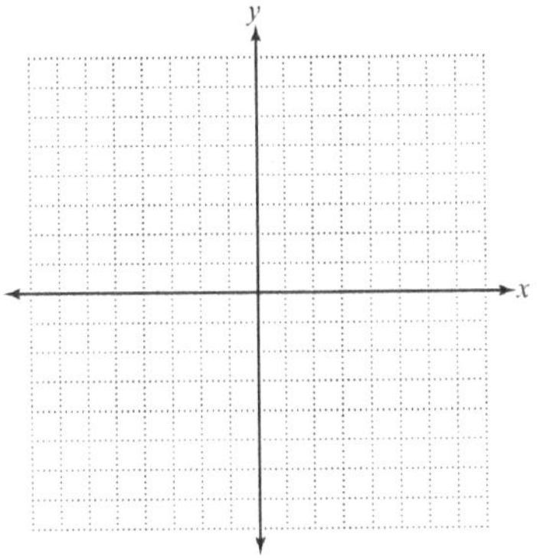

2. $3x + 2y \geq 4$

2.

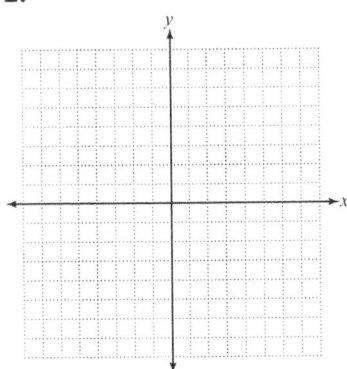

3. $x + 4y \leq -8$

3.

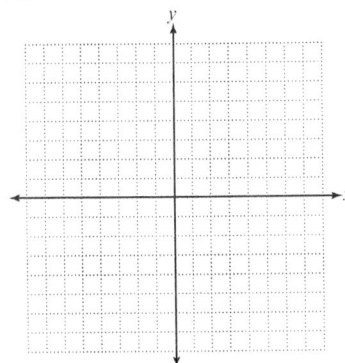

Copyright © 2013 Pearson Education, Inc.   371

Key Terms

Use the vocabulary terms listed below to complete each statement in exercises 1–3.

dashed **system** **solid**

1. A _____ of linear inequalities is two or more linear inequalities with the same variables.

2. If the inequality you are solving is $ax + by \leq c$, draw $ax + by = c$ as a _____ line.

3. If the inequality you are solving is $ax + by > c$, draw $ax + by = c$ as a _____ line.

Practice Problems

Objective A Graphically represent the solutions of a system of linear inequalities.

Graph the solutions of the following systems of linear inequalities.

1. $\begin{cases} x + y < 3 \\ x - y > -1 \end{cases}$

1.

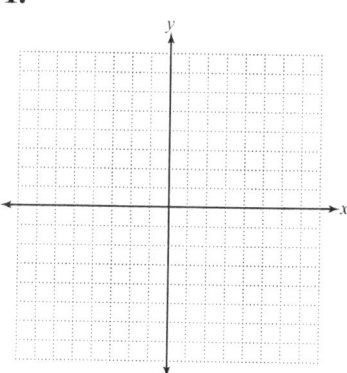

2. $\begin{cases} y - x \geq 2 \\ y + 2x < 4 \end{cases}$

2.

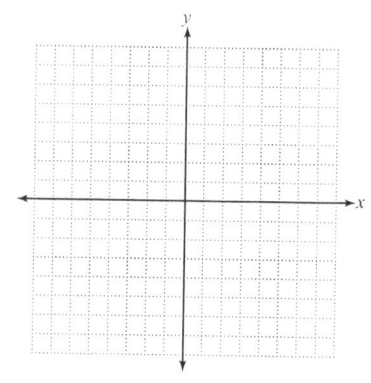

Name:
Instructor:

Date:
Section:

3. $\begin{cases} 4x + y \leq 5 \\ x - 3y < -6 \end{cases}$

3.

4. $\begin{cases} y \geq 3 \\ 2x + 3y < 9 \end{cases}$

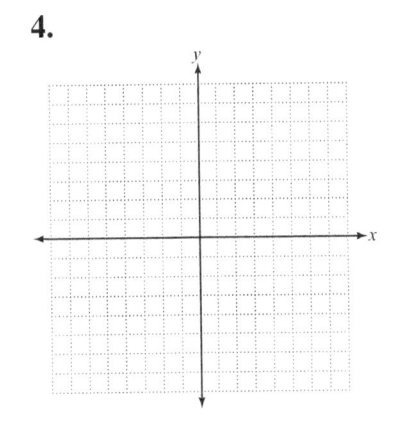

4.

5. $\begin{cases} x \geq 0 \\ x - 2y \leq 4 \end{cases}$

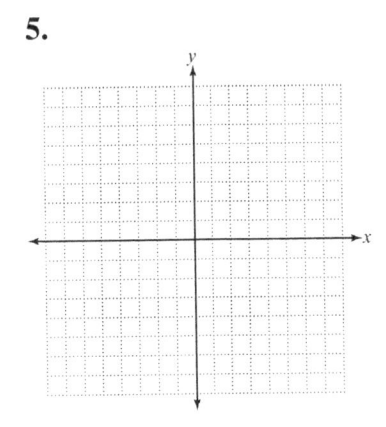

5.

6. $\begin{cases} x+5y>15 \\ -5x+2y\leq 14 \end{cases}$

6.

7. $\begin{cases} y\leq 5 \\ x\geq -3 \end{cases}$

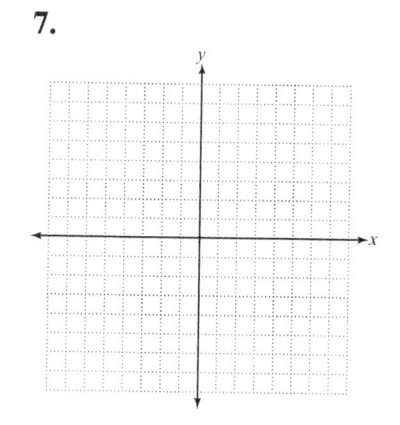

7.

8. $\begin{cases} 3x-4y<-8 \\ x+2y>4 \end{cases}$

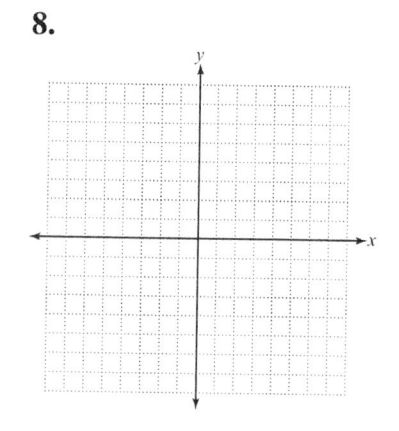

8.

Name:
Instructor:

Date:
Section:

9. $\begin{cases} x > 3 \\ y > -1 \end{cases}$

9.

10. $\begin{cases} y \geq x \\ y \leq 2x+3 \end{cases}$

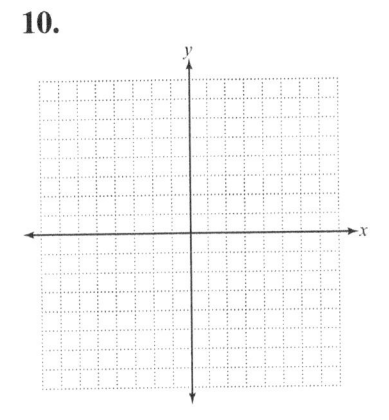

10.

Name: Date:
Instructor: Section:

Chapter 10   ROOTS AND RADICALS

10.1   Defining and Finding Roots

Learning Objectives
A   Find square roots and higher order roots of numbers.
B   Determine whether a root is rational or irrational.
C   Find decimal approximations of square and cube roots, using a calculator or a table.
D   Find roots of variables raised to powers.

Getting Ready

Evaluate the following powers of numbers:

1. $(-4)^2$

2. 3^5

3. $(-3)^3$

4. 6^3

Simplify the following using $(a^m)^n = a^{mn}$:

5. $(x^4)^3$

6. $(a^2)^4$

1._____

2._____

3._____

4._____

5._____

6._____

Key Terms

Use the vocabulary terms listed below to complete each statement in exercises 1–8.

perfect squares	n^{th} root	radicand	root index
$\sqrt{}$	irrational	square root	radical

1. Numbers like $\sqrt{3}$ are called_____.

2. The inverse of squaring a number is finding the _____ .

3. In the notation $\sqrt[n]{a}$, n is called the _____ .

Copyright © 2013 Pearson Education, Inc.   377

4. In the notation $\sqrt[n]{a}$, a is called the _____.

5. In the notation $\sqrt[n]{a}$, the entire expression is called a(n) _____ expression.

6. The _____ of a given number is a number whose n^{th} power equals the given number.

7. Numbers which have rational square roots are called _____.

8. _____ is called a radical sign.

Practice Problems

Objective A Find square roots and higher order roots of numbers.

Find the roots if they exist. If the root does not exist, write, "does not exist as a real number."

1. $\sqrt{36}$ 1._____

2. $-\sqrt{49}$ 2._____

3. $\pm\sqrt{121}$ 3._____

4. $\sqrt{196}$ 4._____

5. $-\sqrt{225}$ 5._____

6. $\pm\sqrt{289}$ 6._____

7. $\sqrt{-36}$ 7._____

8. $-\sqrt{-64}$ 8._____

9. $\sqrt[3]{125}$ 9._____

10. $-\sqrt[3]{8}$ 10._____

Name: Date:
Instructor: Section:

11. $\sqrt[3]{-64}$

11._____

12. $\sqrt[4]{-16}$

12._____

13. $\sqrt[5]{-32}$

13._____

14. $-\sqrt[5]{-32}$

14._____

15. $-\sqrt[3]{-1}$

15._____

16. $\sqrt[4]{-1}$

16._____

Objectives B, C Determine whether a root is rational or irrational. Find decimal approximations of square and cube roots, using a calculator or a table.

Identify each of the numbers as rational or irrational. If the number is rational, find the root exactly. If the number is irrational, approximate the number to the nearest thousandth by using a calculator or Table 1 in your textbook.

17. $\sqrt{28}$

17._____

18. $\sqrt{45}$

18._____

19. $\sqrt{.25}$

19._____

20. $\sqrt[3]{16}$

20._____

21. $\sqrt[3]{36}$

21._____

22. $\sqrt[3]{.008}$

22._____

23. $\sqrt{784}$

23._____

24. $\sqrt[3]{168}$

24._____

Objective D Find roots of variables raised to powers.

Find the roots. Assume that all variables represent nonnegative numbers.

25. $\sqrt{x^8}$ 25._____

26. $\sqrt{y^6}$ 26._____

27. $\sqrt[3]{d^{12}}$ 27._____

28. $\sqrt[5]{m^{20}}$ 28._____

29. $\sqrt[4]{d^4}$ 29._____

30. $\sqrt{x^{14}}$ 30._____

Find the roots. Assume that all expressions have nonnegative values.

31. $\sqrt{(x+7)^2}$ 31._____

32. $\sqrt{(5x-4y)^2}$ 32._____

33. $\sqrt{y^2+10y+25}$ 33._____

34. $\sqrt{16x^2-40xy+25y^2}$ 34._____

Name:
Instructor:
Date:
Section:

Chapter 10 ROOTS AND RADICALS

10.2 Simplifying Radicals

Learning Objectives
A   Simplify square roots by using the product rule.
B   Use the alternative method to simplify radicals.
C   Simplify higher order roots by using the product rule.

Getting Ready

Multiply or divide the following whole numbers:

1. $48 \cdot 2$

    1._____

2. $35 \cdot 4$

    2._____

3. $\dfrac{152}{38}$

    3._____

4. $\dfrac{75}{15}$

    4._____

5. Write 84 in terms of its prime factors.

    5._____

Find the following square roots:

6. $\sqrt{25}$

    6._____

7. $\sqrt{49}$

    7._____

Find the following cube roots:

8. $\sqrt[3]{64}$

    8._____

9. $\sqrt[3]{216}$

    9._____

Simplify the following:

10. $x^{11} \cdot x^3$

10._____

11. $a^3 b^4 \cdot ab^2$

11._____

Key Terms

Use the vocabulary terms listed below to complete each statement in exercises 1–2.

simplified form $\sqrt[n]{a} \cdot \sqrt[n]{b} = \sqrt[n]{a \cdot b}$

1. If $\sqrt[n]{a}$ and $\sqrt[n]{b}$ are both real numbers, then_____.

2. A radical is in _____ if the radicand contains no factor that can be written to a power greater than or equal to the index.

Practice Problems

Objectives A, B   Simplify square roots by using the product rule. Use the alternative method to simplify radicals.

Simplify the following. Assume that all variables have nonnegative values.

1. $\sqrt{24}$

1._____

2. $\sqrt{56}$

2._____

3. $\sqrt{294}$

3._____

4. $3\sqrt{72}$

4._____

5. $7\sqrt{175}$

5._____

6. $\sqrt{x^5}$

6._____

Name:
Instructor:

Date:
Section:

7. $\sqrt{y^{19}}$

7._____

8. $\sqrt{18a^9}$

8._____

9. $\sqrt{x^2 y^7 z^{11}}$

9._____

10. $a^2 b \sqrt{a^3 b^8}$

10._____

11. $p^2 q^3 \sqrt{42 p^3 q}$

11._____

12. $\sqrt{200 x^{13}}$

12._____

13. $2t^3 \sqrt{12 t^5 w^3}$

13._____

14. $8x \sqrt{180 x^4 y^6 z^9}$

14._____

15. $3m^3 n \sqrt{441 m^{16} n^{25}}$

15._____

Objective C Simplify higher order roots by using the product rule.

Simplify the following higher order roots. Assume that all variables have nonnegative values.

16. $\sqrt[3]{56}$

16._____

17. $\sqrt[3]{136}$

17._____

18. $\sqrt[3]{250}$

18._____

19. $\sqrt[3]{x^9}$

19._____

20. $\sqrt[3]{y^{14}}$

20._____

21. $\sqrt[3]{m^{26}}$

21._____

22. $\sqrt[3]{27a^5}$

22._____

23. $\sqrt[3]{64p^7q^{10}}$

23._____

24. $\sqrt[3]{343y^{16}z^2}$

24._____

25. $2b\sqrt[3]{24b^6c^5}$

25._____

26. $-5n^2\sqrt[3]{128m^4n^{11}}$

26._____

27. $p^2q^3\sqrt[3]{p^{21}q^{22}}$

27._____

28. $3x^2y^7\sqrt[3]{-2x^4y^{14}}$

28._____

29. $10\sqrt[3]{500}$

29._____

30. $18\sqrt[3]{189w^{27}}$

30._____

Name:
Instructor:

Date:
Section:

Chapter 10   ROOTS AND RADICALS

10.3   Products and Quotients of Radicals

Learning Objectives
A   Find the products of radicals and simplify the results.
B   Find the quotients of radicals and simplify the results.
C   Simplify radical expressions by using both the product and quotient rules.

Getting Ready

Find the following products or quotients:

1. $4 \cdot 18$

1._____

2. $3x^3 \cdot 7x^5$

2._____

3. $\dfrac{672}{6}$

3._____

4. $\dfrac{64x^3y^5}{4x^2y}$

4._____

Find the following square roots:

5. $\sqrt{81}$

5._____

6. $\sqrt{a^8b^2}$

6._____

Simplify the following radical expressions:

7. $\sqrt{80x^5}$

7._____

8. $\sqrt{450a^3b^6}$

8._____

Copyright © 2013 Pearson Education, Inc.   385

Key Terms

Use the vocabulary terms listed below to complete each statement in exercises 1–3.

$\sqrt[n]{a \cdot b}$ $\sqrt[n]{\dfrac{a}{b}}$ a

1. $\dfrac{\sqrt[n]{a}}{\sqrt[n]{b}} = $ _____.

2. If $\sqrt[n]{a}$ and $\sqrt[n]{b}$ are both real numbers, then $\sqrt[n]{a} \cdot \sqrt[n]{b} = $ _____.

3. If $\sqrt[n]{a}$ is a real number then $\left(\sqrt[n]{a}\right)^n = $ _____.

Practice Problems

Objective A Find the products of radicals and simplify the results.

Find the products or powers of roots. Express all answers in simplified form. Assume that all variables have nonnegative values.

1. $\sqrt{5} \cdot \sqrt{11}$

    1._____

2. $\sqrt{6} \cdot \sqrt{8}$

    2._____

3. $\sqrt{3} \cdot \sqrt{17}$

    3._____

4. $\sqrt{18} \cdot \sqrt{32}$

    4._____

5. $\sqrt{12} \cdot \sqrt{a}$

    5._____

386 Copyright © 2013 Pearson Education, Inc.

Name: Date:
Instructor: Section:

6. $\sqrt{7} \cdot \sqrt{y}$

6._____

7. $\sqrt{x} \cdot \sqrt{x^5}$

7._____

8. $\sqrt{m^3} \cdot \sqrt{m^7}$

8._____

9. $\sqrt{z} \cdot \sqrt{z}$

9._____

10. $2\sqrt{a} \cdot 5\sqrt{a^3}$

10._____

11. $4\sqrt{32} \cdot 2\sqrt{54}$

11._____

12. $3\sqrt{5} \cdot \sqrt{40}$

12._____

13. $\sqrt{p^6} \cdot \sqrt{p^7}$

13._____

14. $\sqrt{a^3b^5} \cdot \sqrt{ab^7}$

14._____

15. $2xy^2\sqrt{x^3y^9} \cdot 3x^3\sqrt{x^7y}$

15._____

16. $6\sqrt{14} \cdot 7\sqrt{22}$

16._____

17. $\left(\sqrt{x^5}\right)^2$

17._____

18. $\left(\sqrt[3]{9y^2}\right)^3$

18._____

19. $\left(\sqrt{z}\right)^2$

19._____

20. $\left(\sqrt[3]{v}\right)^3$

20._____

21. $\left(\sqrt[4]{6p^7}\right)^4$

21._____

22. $\left(\sqrt[6]{18w^2vz^8}\right)^6$

22._____

Name: Date:
Instructor: Section:

Objective B Find the quotients of radicals and simplify the results.

Find the quotients, assuming that all variables have positive values. Express answers in simplest form.

23. $\sqrt{\dfrac{25}{16}}$

23._____

24. $\sqrt{\dfrac{49}{100}}$

24._____

25. $\dfrac{\sqrt{128}}{\sqrt{2}}$

25._____

26. $\dfrac{\sqrt{300}}{\sqrt{3}}$

26._____

27. $\dfrac{2\sqrt{72}}{6\sqrt{2}}$

27._____

28. $\dfrac{10\sqrt{120}}{5\sqrt{12}}$

28._____

29. $\dfrac{\sqrt{a^6 b^5}}{\sqrt{a^3 b}}$

29._____

30. $\dfrac{\sqrt{x^8 y^9}}{\sqrt{x^3 y^6}}$

30._____

31. $\dfrac{\sqrt{18p^9}}{\sqrt{6p^2}}$

31._____

32. $\dfrac{18\sqrt{42y^5 z^{12}}}{3\sqrt{6y^4 z^6}}$

32._____

Objective C Simplify radical expressions by using both the product and quotient rules.

Simplify the following.

33. $\sqrt{\dfrac{14}{11}} \cdot \sqrt{\dfrac{2}{11}}$

33._____

34. $\sqrt{\dfrac{x^5}{3}} \cdot \sqrt{\dfrac{x^7}{3}}$

34._____

35. $\sqrt{\dfrac{6x}{5}} \cdot \sqrt{\dfrac{21}{125}}$

35._____

Name:
Instructor:

Date:
Section:

Chapter 10   ROOTS AND RADICALS

10.4   Addition, Subtraction, and Mixed Operations with Radicals

Learning Objectives
A   Add and subtract like radicals.
B   Simplify radicals and then add and/or subtract like radicals.
C   Simplify radical expressions that contain mixed operations.
D   Find products of radical expressions involving sums or differences.

Getting Ready

Add the following like terms:

1. $6x - 5x$

1._____

2. $4x^3y^2 - 8x^3y^2$

2._____

3. $4\sqrt{75}$

3._____

Multiply or divide the following radical expressions:

4. $\sqrt{5} \cdot \sqrt{30}$

4._____

5. $\dfrac{\sqrt{180}}{\sqrt{12}}$

5._____

6. Simplify $6x(4+7x)$ using the distributive property.

6._____

7. Multiply $(x-7y)(4x+8y)$ using FOIL.

7._____

8. Simplify $(3x+4y)^2$.

8._____

9. Multiply $(x+4y)(x-4y)$.

9._____

10. Reduce $\dfrac{8+16x}{8}$ to lowest terms.

10._____

Copyright © 2013 Pearson Education, Inc.

Key Terms

Use the vocabulary terms listed below to complete each statement in exercises 1–3.

    like radical expressions   **coefficients**   **conjugates**

1. Expressions of the form $a+b$ and $a-b$ are called _____.

2. _____ have the same radicand and the same root indices.

3. To add like radical expressions, add the _____ and leave the radical portion unchanged.

Practice Problems

Objective A Add and subtract like radicals.

Find the sums of radicals. Assume that all variables have nonnegative values.

1. $6\sqrt{5} + 17\sqrt{5}$ 1._____

2. $3\sqrt{7} - 15\sqrt{7}$ 2._____

3. $14\sqrt{17} + 6\sqrt{17}$ 3._____

4. $-8\sqrt{x} - 17\sqrt{x}$ 4._____

5. $3a\sqrt{b} + 2a\sqrt{b}$ 5._____

6. $5q\sqrt[3]{4} - 16q\sqrt[3]{4}$ 6._____

Objective B Simplify radicals and then add and/or subtract like radicals.

Simplify the radicals and then find the sums. Assume that all variables have nonnegative values.

7. $-6\sqrt{6} + 8\sqrt{150}$ 7._____

Name:
Instructor:

Date:
Section:

8. $-7\sqrt{27}+2\sqrt{192}$ 8._____

9. $9\sqrt{7}+2\sqrt{63}$ 9._____

10. $2\sqrt{12}-7\sqrt{75}$ 10._____

11. $-3\sqrt{6}+6\sqrt{24}$ 11._____

12. $-7\sqrt{2}-9\sqrt{18}$ 12._____

13. $3\sqrt{27x^3}+7x\sqrt{192x}$ 13._____

14. $-6\sqrt{128a^5}+2\sqrt{200a^5}$ 14._____

15. $7\sqrt{5p^4}-9p\sqrt{20p^2}$ 15._____

16. $-2\sqrt{108}-2\sqrt{48}+8\sqrt{27}$ 16._____

17. $\sqrt{80}+5\sqrt{180}+9\sqrt{245}$ 17._____

18. $\sqrt{27}+8\sqrt{12}-10\sqrt{243}$ 18._____

Objective C Simplify radical expressions that contain mixed operations.

Simplify the following.

19. $\sqrt{6} \cdot \sqrt{3} - 5 \cdot \sqrt{50}$

19._____

20. $\sqrt{6} \cdot \sqrt{2} - 5 \cdot \sqrt{75}$

20._____

21. $\dfrac{\sqrt{180}}{\sqrt{16}} + \dfrac{\sqrt{45}}{6}$

21._____

22. $\dfrac{4\sqrt{96}}{\sqrt{3}} + 6\sqrt{72}$

22._____

Objective D Find products of radical expressions involving sums or differences.

Simplify the products. Assume that all variables have nonnegative values.

23. $\sqrt{7}\left(6 - \sqrt{7}\right)$

23._____

24. $\sqrt{6}\left(\sqrt{6} - \sqrt{20}\right)$

24._____

25. $8\sqrt{5}\left(4\sqrt{5} + 6\sqrt{3}\right)$

25._____

26. $\left(5 + \sqrt{3}\right)\left(7 - \sqrt{2}\right)$

26._____

Name:
Instructor:

Date:
Section:

27. $(9+\sqrt{x})(8+\sqrt{x})$

27._____

28. $(10-3\sqrt{2})(1+7\sqrt{2})$

28._____

29. $(\sqrt{a}+6\sqrt{b})(\sqrt{a}-3\sqrt{b})$

29._____

30. $(\sqrt{w}+18)(\sqrt{w}-15)$

30._____

31. $(3\sqrt{5}+2\sqrt{6})(8\sqrt{3}-\sqrt{7})$

31._____

32. $(3+\sqrt{11})^2$

32._____

33. $(5-\sqrt{6})^2$

33._____

34. $(1+3\sqrt{7})^2$

34._____

35. $\dfrac{12+2\sqrt{6}}{4}$

35._____

Copyright © 2013 Pearson Education, Inc.

36. $\dfrac{9-18\sqrt{7}}{3}$ 36._____

Find the products of the given conjugates. Assume that all variables have nonnegative values.

37. $\left(6+\sqrt{2}\right)\left(6-\sqrt{2}\right)$ 37._____

38. $\left(8+4\sqrt{3}\right)\left(8-4\sqrt{3}\right)$ 38._____

39. $\left(7+\sqrt{x}\right)\left(7-\sqrt{x}\right)$ 39._____

40. $\left(4\sqrt{2}-5\right)\left(4\sqrt{2}+5\right)$ 40._____

Name: Date:
Instructor: Section:

Chapter 10 ROOTS AND RADICALS

10.5 Rationalizing the Denominator

Learning Objectives
A   Rationalize the denominator when the denominator contains a single radical that is a square root or a cube root.
B   Rationalize the denominator when the radicand is a fraction.
C   Rationalize denominators of the form $a \pm b$, where a and/or b are square roots.

Getting Ready

Multiply the following radical expressions:

1. $\sqrt{2x} \cdot \sqrt{32x^5}$ 1._____

2. $\sqrt[3]{2x} \cdot \sqrt[3]{4x^2}$ 2._____

3. $7(8 - \sqrt{3})$ 3._____

4. $(5 + \sqrt{3})(5 - \sqrt{3})$ 4._____

5. $(\sqrt{a} + \sqrt{b})(1 - \sqrt{b})$ 5._____

Find the following roots:

6. $\sqrt{36x^6}$ 6._____

7. $\sqrt[3]{64x^9}$ 7._____

8. Write $16b^4$ in terms of prime factors.   8._____

9. Simplify $\dfrac{8 + 4\sqrt{3}}{10}$.   9._____

Copyright © 2013 Pearson Education, Inc.

Key Terms

Use the vocabulary terms listed below to complete each statement in exercises 1–3.

        one   conjugates   **rationalizing the denominator**

1. The process of making the denominator of a fraction a rational number is called_____.

2. Any expression can be multiplied by_____ and it will not change the value of that expression.

3. $a + \sqrt{b}$ and $a - \sqrt{b}$ are called_____.

Practice Problems

Objective A Rationalize the denominator when the denominator contains a single radical that is a square root or a cube root.

Rationalize the denominators. Assume that all variables have positive values.

1. $\dfrac{9}{\sqrt{11}}$ 1._____

2. $\dfrac{15}{\sqrt{15}}$ 2._____

3. $\dfrac{-18}{\sqrt{3}}$ 3._____

4. $\dfrac{24}{\sqrt{17}}$ 4._____

5. $\dfrac{16}{\sqrt{15}}$ 5._____

Name: Date:
Instructor: Section:

6. $\dfrac{-a}{\sqrt{7}}$

6._____

7. $\dfrac{2w}{\sqrt{w}}$

7._____

8. $\dfrac{7a}{\sqrt{2}}$

8._____

9. $\dfrac{2}{\sqrt{p}}$

9._____

10. $\dfrac{6a}{\sqrt{5}}$

10._____

11. $\dfrac{\sqrt{5}}{\sqrt{24}}$

11._____

12. $\dfrac{\sqrt{3}}{\sqrt{32}}$

12._____

Objective B Rationalize the denominator when the radicand is a fraction.

Simplify. Assume that all variables have positive values.

13. $\sqrt{\dfrac{49}{2}}$

13._____

Copyright © 2013 Pearson Education, Inc.

14. $\sqrt{\dfrac{16}{3}}$ 14._____

15. $\sqrt{\dfrac{81}{5}}$ 15._____

16. $\sqrt{\dfrac{4}{5p}}$ 16._____

17. $\sqrt{\dfrac{16w^7}{t}}$ 17._____

18. $\sqrt{\dfrac{7}{6p}}$ 18._____

19. $\sqrt{\dfrac{25a^5}{b}}$ 19._____

20. $\sqrt{\dfrac{11}{2m}}$ 20._____

Rationalize the denominator. Assume that all variables have positive values.

21. $\dfrac{\sqrt[3]{7}}{\sqrt[3]{5}}$ 21._____

22. $\dfrac{3}{\sqrt[3]{y}}$ 22._____

Name:
Instructor:

Date:
Section:

23. $\sqrt[3]{\dfrac{7}{3}}$

23._____

24. $\sqrt[3]{\dfrac{4}{9x^2}}$

24._____

25. $\dfrac{\sqrt[3]{4n}}{\sqrt[3]{25m}}$

25._____

26. $\sqrt[3]{\dfrac{5}{16x}}$

26._____

27. $\dfrac{\sqrt[3]{11a}}{\sqrt[3]{81b}}$

27._____

Objective C   Rationalize denominators of the form $a \pm b$, where a and/or b are square roots.

Rationalize the denominators of the following.

28. $\dfrac{7}{9 - \sqrt{3}}$

28._____

29. $\dfrac{\sqrt{3}}{\sqrt{5} + 7}$

29._____

30. $\dfrac{6}{\sqrt{k} - 12}$

30._____

31. $\dfrac{2}{7-\sqrt{2}}$

31._____

32. $\dfrac{\sqrt{5}+5}{\sqrt{7}-7}$

32._____

33. $\dfrac{\sqrt{6}-12}{\sqrt{3}+1}$

33._____

34. $\dfrac{4+\sqrt{p}}{2-\sqrt{p}}$

34._____

35. $\dfrac{\sqrt{m}+\sqrt{n}}{\sqrt{m}-\sqrt{n}}$

35._____

36. $\dfrac{7+\sqrt{3}}{\sqrt{2}}$

36._____

Name:
Instructor:
Date:
Section:

Chapter 10   ROOTS AND RADICALS

10.6   Solving Equations with Radicals

Learning Objectives
A   Solve equations that contain one radical (square root).
B   Solve equations that have two radicals (square roots).

Getting Ready

1. Simplify $\left(\sqrt{4x+5}\right)^2$.

   1._____

2. Simplify $\left(6-\sqrt{x+2}\right)^2$.

   2._____

For exercises 3 and 4, solve the equations:

3. $5x - 6 = 3x + 2$

   3._____

4. $-50 - 6x = x^2 + 6x - 15$

   4._____

5. Find $\sqrt{\dfrac{16}{9}}$.

   5._____

Key Terms

Use the vocabulary terms listed below to complete each statement in exercises 1–3.

        isolate   **extraneous**   **square**

1. When you transform an equation from one type to another type it is possible that solutions of the new equation may not be solutions of the original equation. Such solutions are called_____ solutions and must be discarded.

2. Once you have isolated a square root radical in an equation the next step in solving is to _____ both sides of the equation.

3. The key to solving equations that contain one radical is to _____ the radical on one side of the equation.

Copyright © 2013 Pearson Education, Inc.   **403**

Practice Problems

Objective A   Solve equations that contain one radical (square root).

Solve the following equations.

1. $\sqrt{x+5} = 3$

   1._____

2. $\sqrt{10m-9} = 9$

   2._____

3. $\sqrt{y} + 6 = 5$

   3._____

4. $\sqrt{2k+1} = 3$

   4._____

5. $\sqrt{k} + 9 = 3$

   5._____

6. $\sqrt{5-x} = 3$

   6._____

Name: Date:
Instructor: Section:

7. $\sqrt{6x+1} = 7$ 7._____

8. $\sqrt{z+4} = 8$ 8._____

9. $\sqrt{5x-4} = 4$ 9._____

10. $\sqrt{8t+9} = 6$ 10._____

11. $\sqrt{2q-8} = 2$ 11._____

12. $\sqrt{v+1} = 10$ 12._____

13. $\sqrt{9x+1} - 1 = x$

13._____

14. $\sqrt{6y+1} - y = 1$

14._____

15. $\sqrt{16x+32} - 6 = x$

15._____

16. $\sqrt{9t+4} = t - 2$

16._____

Objective B Solve equations that have two radicals (square roots).

Solve the following equations.

17. $\sqrt{10x-17} = 3\sqrt{x}$

17._____

Name:
Instructor:

Date:
Section:

18. $\sqrt{6y+8} = \sqrt{5y+5}$

18._____

19. $\sqrt{2m+3} - \sqrt{m+1} = 1$

19._____

20. $\sqrt{2q+5} - \sqrt{q-2} = 3$

20._____

21. $\sqrt{x+6} - \sqrt{2-x} = 4$

21._____

22. $\sqrt{9-x} = 3 - \sqrt{x}$

22._____

23. $\sqrt{11y+3} = \sqrt{8y+6}$

23._____

24. $\sqrt{12-n} = \sqrt{n+2}$

24._____

Name:
Instructor:

Date:
Section:

Chapter 10 ROOTS AND RADICALS

10.7 Pythagorean Theorem

Learning Objectives
A   Find the unknown side of a right triangle.
B   Find the length of a diagonal of a rectangle or square.
C   Find the length of the missing side of a rectangle when given the length of a diagonal and the length of one side.
D   Solve application problems involving right triangles.

Getting Ready

Simplify the following:

1. $\left(5\sqrt{6}\right)^2$

1._____

2. $\left(3\sqrt{2}\right)^2$

2._____

3. Solve $36 + a^2 = 81$ for a^2.

3._____

4. Simplify $\sqrt{76}$.

4._____

Key Terms

Use the vocabulary terms listed below to complete each statement in exercises 1–5.

hypotenuse   **right**   **diagonal**   **legs**   **Pythagorean Theorem**

1. A triangle one of whose angles is 90° is called a(n)_____ triangle.

2. _____ states that $a^2 + b^2 = c^2$ where a and b are legs and c is the hypotenuse of a right triangle.

3. In a right triangle, the sides that form the 90° angle are called the _____ of the triangle.

4. In a right triangle, the side opposite the 90° angle is called the _____ of the triangle.

5. A(n)_____ of a square or a rectangle divides the square or rectangle into two right triangles.

Copyright © 2013 Pearson Education, Inc.   409

Practice Problems

Objective A   Find the unknown side of a right triangle.

Find the value of x in each of the following right triangles.

1.

   (Right triangle with legs 7 in. and 3 in., hypotenuse x)

1._____

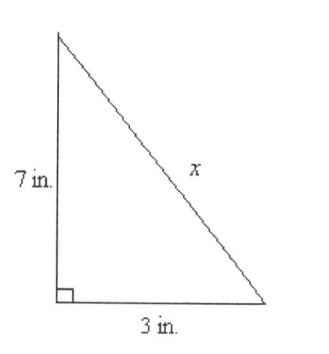

2.

   (Right triangle with legs 5 m and 8 m, hypotenuse x)

2._____

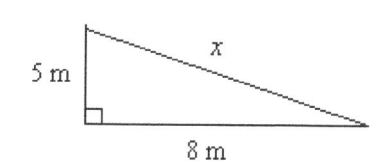

3.

   (Right triangle with leg x, leg 4 cm, hypotenuse 10 cm)

3._____

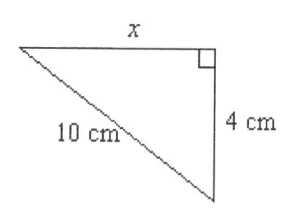

4.

   (Right triangle with legs 7 ft and x, hypotenuse 18 ft)

4._____

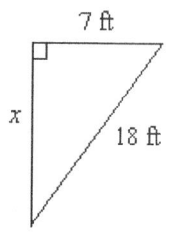

410   Copyright © 2013 Pearson Education, Inc.

Name:
Instructor:
Date:
Section:

Find the length of the unknown side of each of the following right triangles.

5. The lengths of the legs are 8 feet and 12 feet. 5._____

6. The lengths of the legs are 15 meters and 20 meters. 6._____

7. The lengths of the legs are $\sqrt{3}$ inches and 7 inches. 7._____

8. One leg is 2 yards and the hypotenuse is 6 yards. 8._____

9. One leg is 21 feet and the hypotenuse is 35 feet. 9._____

Objective B Find the length of a diagonal of a rectangle or square.

Find the lengths of the diagonals of the following rectangles or squares.

10. 10._____

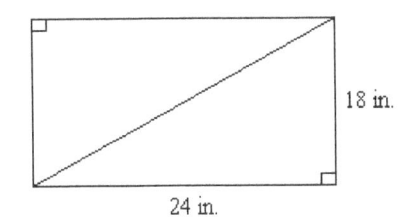

18 in.
24 in.

11.

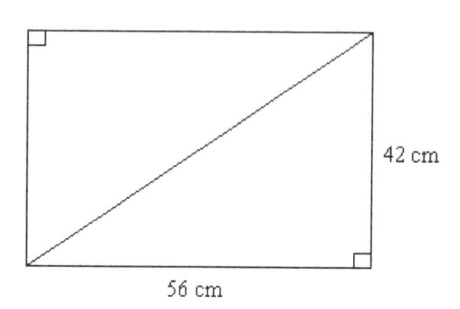

42 cm
56 cm

11._____

12.

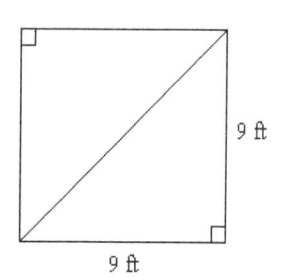

9 ft
9 ft

12._____

13.

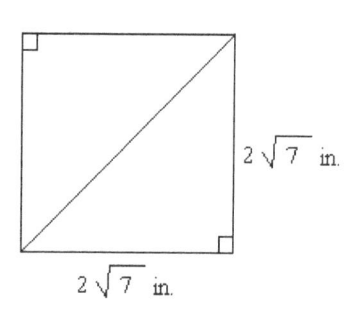

$2\sqrt{7}$ in.
$2\sqrt{7}$ in.

13._____

Objective C Find the length of the missing side of a rectangle when given the length of a diagonal and the length of one side.

Find the length of the unknown side(s) of each of the following.

14.

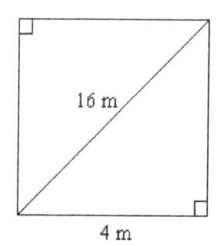

16 m
4 m

14._____

412 Copyright © 2013 Pearson Education, Inc.

Name:
Instructor:

Date:
Section:

15.

18 cm / 12 cm

15. _____

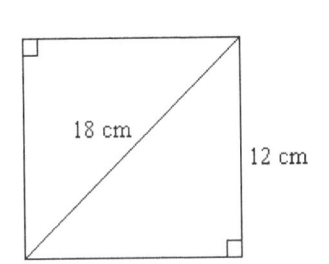

16.

$9\sqrt{2}$ in. / 9 in.

16. _____

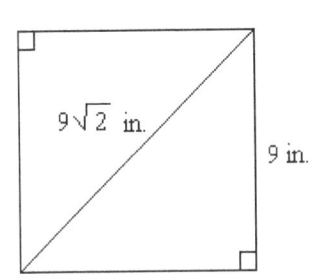

17.

$6\sqrt{6}$ ft / 8 ft

17. _____

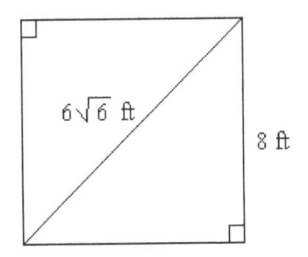

Objective D   Solve application problems involving right triangles.

Solve the following. Round answers to the nearest tenth, if necessary.

18. Kelsey has a rectangular frame that measures 5 feet by 12 feet. She wants to attach a brace on the diagonal for support. What length diagonal brace does she need?

18. _____

Copyright © 2013 Pearson Education, Inc.   413

19. Pam and Mark have a rectangular yard that measures 35 yards by 58 yards. They decided to put a walkway on the diagonal of their yard. What is the length of the walkway?

19._____

20. Martin bought a widescreen television that is advertised as 65 inches. This is the measurement of a diagonal of the screen. If the height of the screen is 36 inches, what is the length of the screen?

20._____

21. There is no straight path for driving from Faisal's house to Chu's house. In fact, it is necessary to travel due east 20 miles, then due north 18 miles. How far apart are their houses?

21._____

22. Bill and Gary leave school at the same time. Bill drives due south at a rate of 40 mph, while Gary drives due west at 55 mph. Find how far apart they are after 1 hour.

22._____

Name: Date:
Instructor: Section:

Chapter 11 SOLVING QUADRATIC EQUATIONS

11.1 Solving Incomplete Quadratic Equations

Learning Objectives
A Solve quadratic equations of the form $ax^2 + bx = 0$.
B Solve quadratic equations of the form $x^2 = k$ by extraction of roots.
C Solve quadratic equations of the form $(ax+b)^2 = c$ by extraction of roots.
D Solve application problems involving incomplete quadratic equations.

Getting Ready

1. Factor $16x^2 - 2x$.

   1._____

2. Solve $6x - 4 = 4$.

   2._____

3. Find $\sqrt{25}$.

   3._____

4. Simplify $\sqrt{48}$.

   4._____

5. Rationalize the denominator of $\dfrac{\sqrt{2}}{\sqrt{5}}$.

   5._____

6. Simplify $\left(-5\sqrt{5}\right)^2$.

   6._____

Key Terms
Use the vocabulary terms listed below to complete each statement in exercises 1–3.

   incomplete quadratic equation   **factorable**   **positive**

1. The symbol $\sqrt{}$ means find the _____ square root only.

2. $ax^2 + bx + c = 0$, in which either b or $c = 0$, is called a(n)_____.

Copyright © 2013 Pearson Education, Inc. 415

3. Not all quadratic polynomials are _____.

Practice Problems

Objective A Solve quadratic equations of the form $ax^2 + bx = 0$.

Solve the following equations. If there are no real solutions, write "no real-number solutions."

1. $3x^2 + 18x = 0$ 1._____

2. $4x^2 - 24x = 0$ 2._____

3. $7y^2 - 28y = 0$ 3._____

4. $10q^2 = 20q$ 4._____

5. $4n^2 = 25n$ 5._____

6. $p^2 - 15p = 0$ 6._____

7. $12x - x^2 = 0$ 7._____

8. $-t^2 = 3t$ 8._____

Name:　　　　　　　　　　　　　　Date:
Instructor:　　　　　　　　　　　Section:

Objective B Solve quadratic equations of the form $x^2 = k$ by extraction of roots.

Solve the following equations. If there are no real solutions, write "no real-number solutions."

9. $6y^2 = 96$

9._____

10. $8m^2 = 88$

10._____

11. $t^2 = -49$

11._____

12. $2p^2 = 200$

12._____

13. $t^2 + 3 = 199$

13._____

14. $-3n^2 + 4 = 9$

14._____

15. $3m^2 = 90$

15._____

16. $y^2 = 18$

16._____

17. $6z^2 + 4 = 388$

17._____

18. $b^2 + 69 = 5$

18._____

19. $3x^2 + 9 = 18$

19._____

20. $2 - 9c^2 = -18$

20._____

Objective C Solve quadratic equations of the form $(ax+b)^2 = c$ by extraction of roots.

Solve the following equations. If there are no real solutions, write "no real-number solutions."

21. $(x-5)^2 = 25$

21._____

22. $(x+8)^2 = 44$

22._____

23. $(2d-3)^2 = 81$

23._____

24. $(5q-1)^2 = \dfrac{9}{25}$

24._____

Name:
Instructor:

Date:
Section:

25. $(a+7)^2 = 15$

25._____

26. $(q-6)^2 + 3 = 0$

26._____

27. $(p+9)^2 - 3 = 0$

27._____

28. $(3y+5)^2 = 24$

28._____

29. $(2x-1)^2 = 42$

29._____

30. $(4b+3)^2 = 121$

30._____

Objective D Solve application problems involving incomplete quadratic equations.

Solve the following application problems.

31. The square of 5 more than a number is 49. Find the number(s).

31._____

32. The square of 6 less than 7 times a number is 144. Find the number(s).

32._____

33. Gloria has a square rug whose area is 225 square feet. Find the length of each side of her rug.

33._____

34. If the length of each side of a square is doubled and then decreased by 3, the area would be 441 square meters. Find the length of each side of the square.

34._____

Name:
Instructor:

Date:
Section:

Chapter 11 SOLVING QUADRATIC EQUATIONS

11.2 Solving Quadratic Equations by Completing the Square

Learning Objectives
A Determine the number to be added to a polynomial of the form $x^2 + bx$ in order to form a perfect-square trinomial.
B Solve quadratic equations by completing the square.

Getting Ready

Simplify the following:

1. $\left(3 + \sqrt{3}\right)^2$ 1._____

2. $\left(x + \dfrac{5}{2}\right)^2$ 2._____

3. $\left(\dfrac{8}{7}\right)^2$ 3._____

4. Factor $a^2 + 9a + \dfrac{81}{4}$.   4._____

5. Simplify $\dfrac{3}{4} \cdot \dfrac{7}{11}$.   5._____

For exercises 6 and 7, solve the equations:

6. $\left(x + \dfrac{7}{2}\right)^2 = \dfrac{25}{4}$ 6._____

7. $(x - 3)^2 = 3$ 7._____

8. Find $\sqrt{\dfrac{25}{16}}$.   8._____

Copyright © 2013 Pearson Education, Inc.   421

9. Simplify $\sqrt{80}$.

9._____

10. Add $-\dfrac{5}{3}+\dfrac{4\sqrt{3}}{3}$.

10._____

11. Combine by putting over a common denominator:
    $-3-\dfrac{\sqrt{5}}{2}$.

11._____

12. Rationalize the denominator: $\sqrt{\dfrac{3}{5}}$.

12._____

Key Terms

Use the vocabulary terms listed below to complete each statement in exercises 1–4.

$\quad$ 1 $\qquad$ $\dfrac{1}{2}b$ $\qquad$ **perfect square trinomial** $\qquad$ **completing the square**

1. A(n) _____ is a trinomial that is the square of a binomial.

2. To find the number needed to make a polynomial of the form x^2+bx a perfect-square trinomial, take _____ and square it.

3. The procedure for finding the number that will make a polynomial of the form x^2+bx a perfect-square trinomial is called _____.

4. In the process of completing the square, if the coefficient of the x^2 term is not _____ divide by that coefficient.

Practice Problems

Objective A Determine the number to be added to a polynomial of the form x^2+bx in order to form a perfect-square trinomial.

Find the number needed to make each polynomial a perfect-square trinomial. Then, write each trinomial as the square of a binomial.

1. x^2-6x+____

1._____

Name:
Instructor:
Date:
Section:

2. $x^2 + 8x + \underline{}$

2._____

3. $x^2 - 12x + \underline{}$

3._____

4. $y^2 + 9y + \underline{}$

4._____

5. $q^2 + 11q + \underline{}$

5._____

6. $m^2 + \dfrac{1}{3}m + \underline{}$

6._____

7. $p^2 - \dfrac{2}{11}p + \underline{}$

7._____

8. $x^2 + \dfrac{2}{5}x + \underline{}$

8._____

Objective B Solve quadratic equations by completing the square.

Solve each equation by completing the square. If the equation has no real solutions, write "no real-number solutions."

9. $m^2 + 12m = -20$

9._____

10. $n^2 + 3n - 9 = 0$

10._____

11. $x^2 + 5x - 5 = 0$ 11._____

12. $y^2 - 18y + 8 = 0$ 12._____

13. $d^2 + 8d + 1 = 0$ 13._____

14. $2x^2 - 7x - 9 = 0$ 14._____

15. $6y^2 + 3y - 3 = 0$ 15._____

16. $-w^2 + 2w = -6$ 16._____

17. $4x^2 + 6x + 48 = 0$ 17._____

18. $3u^2 = -9u + 24$ 18._____

19. $5y^2 + 2y - 6 = 0$ 19._____

20. $2n^2 - 7n + 20 = 0$ 20._____

Name:
Instructor:

Date:
Section:

Chapter 11 SOLVING QUADRATIC EQUATIONS

11.3 Solving Quadratic Equations by the Quadratic Formula

Learning Objectives
A   Solve quadratic equations by the quadratic formula.

Getting Ready

1. Solve $x^2 + 8x - 2 = 0$ by completing the square.

   1._____

Given $\dfrac{-b \pm \sqrt{b^2 - 4ac}}{2a}$, evaluate for the following values of a, b, and c:

2. $a = 2$, $b = -4$, $c = 1$

   2._____

3. $a = 3$, $b = -5$, $c = 2$

   3._____

4. Simplify $\sqrt{350}$.

   4._____

5. Reduce $\dfrac{-8 + 4\sqrt{17}}{4}$ to lowest terms.

   5._____

Key Terms

Use the vocabulary terms listed below to complete each statement in exercises 1–4.

$ax^2 + bx + c = 0$ Quadratic Formula   quadratic equation   coefficients

1. The solutions of a quadratic equation are determined by its _____.

2. Any _____ can be solved using the Quadratic Formula.

3. The _____ says that the solutions of $ax^2 + bx + c = 0$ are $x = \dfrac{-b \pm \sqrt{b^2 - 4ac}}{2a}$.

4. To identify a, b, and c correctly for use in the Quadratic Formula, the quadratic equation must be written in the form _____.

Copyright © 2013 Pearson Education, Inc.

Practice Problems

Objective A Solve quadratic equations by the quadratic formula.

Solve the following using the quadratic formula. If the equation has no real solutions, write "no real-number solutions."

1. $x^2 - 5x + 14 = 0$ 1._____

2. $y^2 + 3y - 8 = 0$ 2._____

3. $m^2 = 4m + 2$ 3._____

4. $n^2 - 1 = 7n$ 4._____

Name:
Instructor:

Date:
Section:

5. $2t^2 + 8t - 5 = 0$

5._____

6. $3q^2 - 7q + 9 = 0$

6._____

7. $5x^2 = 6x - 1$

7._____

8. $2 - y = 9y^2$

8._____

9. $4d^2 + 12d + 9 = 0$

9._____

10. $6w^2 + 23w = -20$ 10._____

11. $2x^2 - 7x - 9 = 0$ 11._____

12. $10x^2 + x = 1$ 12._____

13. $2n^2 + 10n + 5 = 0$ 13._____

14. $10x^2 = -2x - 11$ 14._____

Name:
Instructor:

Date:
Section:

15. $3q^2 + 6q + 2 = 0$

15._____

16. $5r^2 = -8r - 2$

16._____

17. $3y^2 + 10y = -4$

17._____

18. $6x^2 = 4x - 3$

18._____

19. $3q^2 + 2q - 5 = 0$

19._____

20. $8y^2 - 5 = 2y$ 20._____

Name: Date:
Instructor: Section:

Chapter 11 SOLVING QUADRATIC EQUATIONS

11.4 Quadratic Equations with Complex Solutions

Learning Objectives
A Write complex numbers in terms of i.
B Add and subtract complex numbers.
C Multiply and divide complex numbers.
D Solve quadratic equations that have complex solutions.

Getting Ready

1. Simplify $\sqrt{48}$.

   1._____

2. Simplify $5 - 5i - 3 + 8i$ by adding like terms.

   2._____

3. Simplify $3i(6 - 8i)$ by using the distributive property.

   3._____

Multiply the following binomials:

4. $(5 + 4i)(3 - 2i)$

   4._____

5. $(8 + i)(8 - i)$

   5._____

6. Rewrite $\dfrac{7 - 4i}{5}$ using $\dfrac{a+b}{c} = \dfrac{a}{c} + \dfrac{b}{c}$.

   6._____

7. Solve $(x - 8)^2 = 25$ by using extraction of roots.

   7._____

8. Solve $x^2 - 3x - 8 = 0$ by using the quadratic formula.

   8._____

Key Terms

Use the vocabulary terms listed below to complete each statement in exercises 1–5.

      real number   **complex number**   **imaginary unit**
      complex conjugates   **standard form**

1. $a + bi$ and $a - bi$ are _____.

2. The _____, i, is the number whose square is -1.

3. The product of complex conjugates is a(n) _____.

4. When a complex number is written in the form $a+bi$ it is said to be in _____.

5. A(n) _____ is any number that can be written in the form of $a+bi$, where a and b are real numbers.

Practice Problems

Objective A Write complex numbers in terms of *i*.

Write each of the following as the product of a real number and i.

1. $\sqrt{-121}$ 1._____

2. $\sqrt{-144}$ 2._____

3. $\sqrt{-8}$ 3._____

4. $\sqrt{-24}$ 4._____

5. $\sqrt{-17}$ 5._____

6. $\sqrt{-31}$ 6._____

Write each of the following as a complex number in standard form $a+bi$.

7. 18   7._____

8. -21 8._____

9. $12i$ 9._____

Name: Date:
Instructor: Section:

10. $-5i$ 10._____

11. $\sqrt{32}$ 11._____

12. $\sqrt{-112}$ 12._____

13. $5+\sqrt{-25}$ 13._____

14. $-7-\sqrt{-48}$ 14._____

Objective B   Add and subtract complex numbers.

Find the sums or differences of the complex numbers. Express answers in standard form $a+bi$.

15. $16+(8-3i)$ 15._____

16. $7i+(5+2i)$ 16._____

17. $(1+i)+(3-7i)$ 17._____

18. $(2+3i)-(4-9i)$ 18._____

19. $(-6+12i)-(18+7i)$ 19._____

20. $(-4-3i)-(10-3i)$ 20._____

Copyright © 2013 Pearson Education, Inc.

21. $(2+9i)+(-2+17i)$

21._____

22. $(4+i)-(3-i)$

22._____

Objective C Multiply and divide complex numbers.

Find the products or quotients of the complex numbers. Express answers in standard form $a+bi$.

23. $2i(7+9i)$

23._____

24. $-3i(5-7i)$

24._____

25. $(1+i)(5-i)$

25._____

26. $(2+4i)(-3+2i)$

26._____

27. $(8-i)(8+i)$

27._____

28. $(3+9i)^2$

28._____

29. $\dfrac{2+i}{1-i}$

29._____

Copyright © 2013 Pearson Education, Inc.

Name: Date:
Instructor: Section:

30. $\dfrac{5+6i}{4-5i}$

30._____

31. $\dfrac{6-5i}{2+i}$

31._____

32. $\dfrac{7+i}{8-i}$

32._____

Objective D Solve quadratic equations that have complex solutions.

Find the solutions of the quadratic equations. Express answers in standard form.

33. $(x-3)^2 = -49$

33._____

34. $(x+5)^2 = -100$

34._____

35. $x^2 - 6x + 13 = 0$

35._____

36. $x^2 + 8x + 25 = 0$ **36.** _____

37. $x^2 - x + 1 = 0$ **37.** _____

38. $x^2 + 4x + 5 = 0$ **38.** _____

Name: Date:
Instructor: Section:

Chapter 11   SOLVING QUADRATIC EQUATIONS

11.5   Applications Involving Quadratic Equations

Learning Objectives
Solve the following types of application problems involving quadratic equations:
A   Numbers
B   Geometric figures
C   Pythagorean theorem
D   Distance, rate, and time
E   Work
F   Applications from science

Getting Ready

Solve the following equations:

1. $6x - 14 = 0$

    1._____

2. $3x^2 + 3x + 78 = 204$

    2._____

3. $\dfrac{100}{x} = \dfrac{100}{x+1} + 5$

    3._____

4. The shorter leg of a right triangle is one meter shorter than the longer leg, and the hypotenuse is 5 meters. Find the lengths of the two legs.

    4._____

5. Represent three consecutive odd integers, using x as the variable.

    5._____

Key Terms

Use the vocabulary terms listed below to complete each statement in exercises 1–2.

    original problem   **physical conditions**

1. When solving an application involving a quadratic equation it is possible that a solution of the equation may not satisfy the _____ of the problem.

2. When solving an application problem always check solutions of the equation involved against the wording of the _____.

Copyright © 2013 Pearson Education, Inc.   437

Practice Problems

Objectives A-F Solve application problems involving quadratic equations.

Solve the following.

1. The difference of the reciprocals of two consecutive integers is $\frac{1}{90}$. Find the integers.

   1._____

2. A square sheet of paper measures 20 centimeters on each side. What is the length of the diagonal of this paper?

   2._____

3. A ladder that is 13 feet long reaches 12 feet up a wall. How far is the foot of the ladder from the base of the wall?

   3._____

Name: Date:
Instructor: Section:

4. The length of a rectangular room is 4 feet longer than its width. If the area of the room is 60 square feet, find its dimensions.

4._____

5. Timmy can row 6 miles upstream against a 2 mph current in one hour more than he can row 7 miles downstream. What is his average speed in still water?

5._____

6. The hypotenuse of a right triangle is 7 feet long. One leg is 1.4 feet longer than the other leg. Find the lengths of the legs.

6._____

7. It takes one machine 4 minutes longer than a second machine to sort an order of bolts by different sizes. Together, they can sort the bolts in 4.8 minutes. How long does it take each machine to sort the bolts when working alone?

7._____

8. Betty and Joe leave their homes at 9:00 a.m., going toward each other along the same route, which is 9 miles long. Joe walks at 4 mph while Betty jogs at 8 mph. At what time will they meet?

8._____

9. Suppose a ball is thrown up into the air and its height, in feet, above the ground t seconds after it is thrown is given by $s = -16t^2 + 80t + 10$. How high is the ball after 2 seconds?

9._____

Name: Date:
Instructor: Section:

10. If an object is dropped from a height of 1000 feet and its height s after t seconds is $s = 1000 - 16t^2$, how long will it take for the object to reach a height of 700 feet?

10._____

11. Kyle and Amanda can paint a house together in 6 days. Painting alone, it takes Amanda 9 days longer than it takes Kyle. How long would it take each person painting alone?

11._____

12. The daily profit made by JZ Auto manufacturers is $p = -45x^2 + 2250x - 18000$ where x is the number of cars produced per day. How many cars must be produced per day for the company to make a profit of $9000?

12._____

13. The sum of the squares of two consecutive even integers is 164. Find the integers.

13._____

14. The square of a number is 20 more than the number. Find the number(s).

14._____

15. The length of a rectangular picture is 7 inches more than the width. If the area is 78 square inches, what are the dimensions of the picture?

15._____

Answers to Worksheets for Classroom or Lab Practice

Chapter R BASIC IDEAS

R.1 Reading and Writing Numerals

Key Terms
1. digits   2. numeral   3. place value   4. expanded notation
5. word name

Objective A
1. tens   3. millions   5. ones   7. $8 \cdot 100 = 800$
9. $1 \cdot 1000 = 1000$

Objective B
11. $200 + 60 + 3$ 13. $1,000,000 + 400,000 + 70,000 + 1000 + 200 + 60 + 5$

Objective C
15. eighteen   17. three hundred ninety-eight thousand, four hundred ten
19. five hundred sixty-two   21. 6487   23. 840   25. 182,704,009

R.2 Addition and Subtraction of Whole Numbers

Getting Ready
1. 6422   2. 780   3. 2,919,041

Key Terms
1. variable   2. zero   3. addition   4. subtraction
5. addends   6. sum   7. minuend   8. subtrahend
9. difference

Objectives A, B, C
1. 683   3. 77   5. 1616   7. 22,466   9. 491
11. 1007

Objectives D, E, F
13. 843   15. 75,113   17. 24   19. 23   21. 87

Objective G
23. 11 students

R.3 Multiplication and Division of Whole Numbers

Getting Ready
1. 981   2. 256   3. 57   4. 336

Key Terms
1. factors   2. dividend   3. multiples   4. divisor   5. product

Copyright © 2013 Pearson Education, Inc.   443

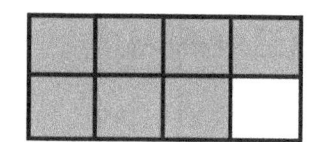

Answers to Worksheets for Classroom or Lab Practice

6. quotient

Objectives A, B, C
1. 456
3. 1233
5. 4644
7. 16,288
9. 128,832
11. 704 bowling pins

Objectives D, E, F
13. 12 R 3
15. 5 R 2
17. 85 R 2
19. 11 R 5
21. 10 R 346
23. 7 trips
25. $28

R.4 A Brief Introduction to Fractions

Getting Ready
1. 17
2. 25
3. 56
4. 7
5a. difference
b. product
c. quotient
d. sum

Key Terms
1. fraction
2. numerator
3. denominator
4. multiplicative inverses
5. lowest terms
6. divide

Objective A
1. $\dfrac{2}{5}$

Objective B
3.

Objective C
5. $4\dfrac{5}{9}$
7. $5\dfrac{2}{7}$
9. $\dfrac{13}{5}$
11. $\dfrac{89}{6}$

Objective D
13. $\dfrac{2}{7}$
15. $\dfrac{7}{10}$

Objective E
17. $\dfrac{1}{18}$
19. $\dfrac{6}{85}$
21. 2

Objective F
23. 36
25. $\dfrac{2}{33}$
27. $\dfrac{21}{40}$
29. $\dfrac{21}{2}$

Answers to Worksheets for Classroom or Lab Practice

Objective G

31. $\dfrac{5}{9}$ 33. $\dfrac{5}{7}$ 35. $\dfrac{13}{21}$ 37. $\dfrac{3}{14}$ 39. $\dfrac{21}{4}$ cups

R.5 Addition and Subtraction of Decimal Numerals

Getting Ready
1. 1093   2. 1474   3. 2264   4. $\dfrac{1}{10{,}000}$ 5. 100
6. 1: sum; 2: difference; 3: sum; 4: product; 5: quotient   7. 25 minutes

Key Terms
1. decimal mixed numerals   2. place values   3. decimal point

Objective A
1. fifty-one hundredths   3. seventy-eight and three hundred five thousandths
5. tenths

Objective B
7. .328   9. 2.06009

Objective C
11. 1.2   13. 9.632   15. 1034.001   17. 5.2829

Objective D
19. .47   21. 1.671   23. 214.0844
25. two hundred forty-seven and $\dfrac{38}{100}$ dollars   27. $195.15

R.6 Multiplication and Division of Decimal Numerals

Getting Ready
1. 612   2. 3174   3. 18   4. 19   5. $\dfrac{1}{1{,}000{,}000}$

Key Terms
1. sum   2. whole number   3. rounding

Objective A
1. 14.5   3. .24   5. 2.6568   7. .24

Objective B
9. 7   11. 1.1

Copyright © 2013 Pearson Education, Inc.

Answers to Worksheets for Classroom or Lab Practice

Objective C
13. .64 15. 1.001

Objective D
17. 1.83 19. 66.7

Objective E
21. .2 23. .42 25. .56 27. $30.50 29. $4719

R.7 Linear Measurement in the American and Metric Systems

Getting Ready
1. 57.6 2. 5190 3. .0617 4. 8.296 5. 120
6. 18 7. 3 8. 280.8

Key Terms
1. meter 2. metric 3. unit-cancellation 4. American

Objective A
1. 60 3. 69 5. 14,960 7. 41

Objective B
9. 26 11. 26,180 13. 93,000 15. .14

Chapter 1 ADDING AND SUBTRACTING INTEGERS AND POLYNOMIALS

1.1 Variables, Exponents, and Order of Operations

Getting Ready
1. 73 2. 26 3. 512 4. 5

Key Terms
1. π 2. variable 3. base 4. exponent 5. constant

Objective A
1. constant: 9; variable: x 3. constant: 1; variable: t

Objective B
5. seven cubed; 343 7. $x \cdot x \cdot x \cdot x \cdot x \cdot x$ 9. $3 \cdot a \cdot a \cdot b \cdot b \cdot b \cdot b$
11. x^3 13. $7c^3 d^2$

Objective C
15. 7 17. 21 19. 576 21. 7 23. 162 25. 1

Answers to Worksheets for Classroom or Lab Practice

Objective D
27. 33   29. 800   31. $240 + 52(12) = \$864$

1.2 Perimeters of Geometric Figures

Getting Ready
1. 75   2. 75.36   3. 176   4. 33   5. 308   6. 34

Key Terms
1. perimeter   2. circumference   3. diameter   4. π

Objective A
1. 15 in.   3. 26 m   5. 34 yd   7. 12.8 ft   9. 17.5 m
11. 52 ft   13. 18π yd; 56.52 yd   15. 34π in.; 106.76 in.

Objective B
17. 18.84 in.   19. 803.84 in.

1.3 Areas of Geometric Figures

Getting Ready
1. 51   2. 47.1   3. 11.5   4. 14.91   5. 12.56   6. 81

Key Terms
1. area   2. square unit   3. right   4. $A = \dfrac{bh}{2}$
5. $A = s^2$ 6. $A = \dfrac{h(B+b)}{2}$ 7. $A = LW$ 8. $A = \pi r^2$

Objective A
1. 36 m^2 3. 60 yd^2 5. 10.24 ft^2 7. 47.5 ft^2 9. 192 ft^2
11. 81π yd^2; 254.3 yd^2 13. 289π in.^{2}; 907.5 in.^{2} 15. 114 cm^2
17. 50.53 ft^2

Objective B
19. 28.26 ft^2

1.4 Volumes and Surface Areas of Geometric Figures

Getting Ready
1. 702   2. 188.4   3. 62.8   4. 904.32

Key Terms
1. volume   2. cube   3. $V = \pi r^2 h$ 4. $V = LWH$

Answers to Worksheets for Classroom or Lab Practice

5. sphere   6. $V = e^3$ 7. $V = \frac{4}{3}\pi r^3$ 8. surface area

Objectives A, B
1. $V = 216$ in.^{3}, $SA = 228$ in.^{2}
3. $V = 343$ mi^3, $SA = 294$ mi^2
5. $V = 4408.56$ cm^3, $SA = 1695.6$ cm^2
7. $V = 1582.56$ cm^3, $SA = 753.6$ cm^2
9. 5572.5 in.^{3}

Objective C
11. 12 ft^3 13. 112 ft^2

1.5 Introduction to Integers

Key Terms
1. natural   2. whole   3. integers   4. opposites
5. graphing   6. absolute value

Objective A
1. 75 miles south

Objective B
3.

Objective C
5. <   7. <   9. =   11. >

Objective D
13. −21   15. −8   17. 7

Objective E
19. 1   21. 24   23. 5

1.6 Addition of Integers

Getting Ready
1. 25   2. 7   3. 4.5   4.

Key Terms
1. associative   2. commutative   3. the same sign
4. opposite signs   5. sum   6. additive inverses
7. additive identity

Objective A
1. 26   3. −31   5. −30   7. .9   9. 15   11. −17

Answers to Worksheets for Classroom or Lab Practice

13. −41

Objective B
15. 6   17. 5   19. $1037

Objective C
21. commutative property of addition
23. additive identity
25. commutative property of addition
27. $12+(-9)$
29. 0

1.7 Subtraction of Integers and Combining Like Terms

Getting Ready
1. 8.1   2. 8   3. 3   4. −5   5. −15   6. 7

Key Terms
1. opposite   2. term   3. distributive property
4. numerical coefficient   5. variable

Objective A
1. 6   3. −9   5. 47   7. 1   9. −31   11. −5.3   13. 3.5
15. −48   17. −2

Objective B
19. $18x$ 21. $-7mn+8$ 23. $12x^2-22y^2+10$
25. $19a^2b-4ab^2+19b$ 27. −11   29. $13ab$

1.8 Polynomial Definitions and Combining Polynomials

Getting Ready
1. 23   2. 11   3. $3a-1$ 4. $13p^2+6p+4$ 5. $-9x$
6. $-2a$

Key Terms
1. algebraic expression   2. term   3. binomial   4. monomial
5. trinomial   6. polynomial   7. degree

Objective A
1. 2, binomial   3. 1, monomial   5. 4

Objective B
7. coefficient: −8; variable: x; degree: 5
9. coefficient: −1; variable: y, z; degree: 16

Copyright © 2013 Pearson Education, Inc.   **449**

Answers to Worksheets for Classroom or Lab Practice

Objective C
11. $-2x^3 + 5x$; degree: 3

Objective D
13. 17   15. $9x+9y$ 17. $-6m-18n$ 19. $18p^2q^2 - 2p^2q - 16pq^2 - 11p$
21. $7y+9$

Chapter 2   LAWS OF EXPONENTS, PRODUCTS AND QUOTIENTS OF INTEGERS AND POLYNOMIALS

2.1   Multiplication of Integers

Getting Ready
1. 96   2. 420   3. 243   4. 64   5. 80   6. 256

Key Terms
1. product   2. commutative property of multiplication   3. negative
4. associative property of multiplication   5. identity   6. positive
7. inverse

Objective A
1. −18   3. −80   5. −120

Objective B
7. 9   9. −216   11. 900   13. 120   15. −11   17. 10

Objective C
19. inverse for multiplication   21. $[-4(5)](-12)$ 23. −10

2.2   Multiplication Laws of Exponents

Getting Ready
1. 14   2. 15   3. 16   4. 36   5. $12a^2$ 6. $8x^3y$

Key Terms
1. factor   2. add   3. multiply

Objective A
1. t^{11} 3. a^{22} 5. $34y^3$

Objective B
7. 3^4d^4 9. $2^4a^4b^4c^4$

Answers to Worksheets for Classroom or Lab Practice

Objective C
11. m^{60}
13. x^{32}

Objective D
15. $5^3 d^6$
17. $2^8 x^6$
19. $144 p^{14} q^{14}$
21. $7^{18} m^{36}$
23. $-9^4 b^{10}$
25. $-5^{12} x^{42}$
27. $-2x^2$
29. $-508 n^6$

2.3 Products of Polynomials

Getting Ready
1. 24
2. $-12 x^4 y^6$
3. -40
4. $5x^2 + 2x + 7$
5. 77 square inches

Key Terms
1. triangle
2. distributive property
3. rectangle

Objective A
1. $14a - 21b$
3. $8m - 20n$
5. $-12 y^7 + 48 y^4$
7. $2t^7 - 4t^5 - 9t$
9. $18 a^5 b^{10} + 10 a^7 b^{12}$
11. $15x - 9$
13. $6y^2 + 8y - 35$
15. $10m^2 - 17mn - 12n^2$
17. $24a^2 - 7a - 10$

Objective B
19. $y^2 + 10y + 21$
21. $w^2 - 17w + 72$
23. $10b^2 - 3b - 4$
25. $b^2 + b - 30$
27. $2x^3 - x^2 - 10x - 7$
29. $16x^2 - 2x - 3$
31. $3x^2 + 11x + 10$
33. $12x^2$

2.4 Special Products

Getting Ready
1. $4y^2 + 5y$
2. $3x$
3. $-42 c^2$
4. 15
5. 36

Key Terms
1. conjugate pairs
2. difference of squares
3. FOIL

Objective A
1. $x^2 + 11x + 28$
3. $w^2 - w - 12$
5. $18b^2 - 9b - 20$
7. $6x^2 + 13xy + 6y^2$

Objective B
9. $x^2 - 64$
11. $16 z^2 - 25$
13. $36 c^2 - 49 d^2$

Objective C
15. $x^2 + 22x + 121$
17. $4w^2 - 20w + 25$
19. $16x^2 + 40xy + 25 y^2$

Copyright © 2013 Pearson Education, Inc.   451

Answers to Worksheets for Classroom or Lab Practice

21. $36a^4 + 36a^2b + 9b^2$ 23. $x^2 - 196$ 25. $36a^2 - 12ab + b^2$
27. $14m^2 - 53mn + 14n^2$

Objective D
29. $a^3 + 1$ 31. $a^3 - 343$ 33. $8a^3 + 125b^3$

2.5 Division of Integers and Order of Operations with Integers

Getting Ready
1. 9   2. 13   3. -48 4. 128   5. 42   6. 38
7. 26

Key Terms
1. negative   2. zero   3. quotient   4. positive
5. multiplication

Objective A
1. -3 3. -3 5. 0   7. -21 9. -2 11. -12

Objective B
13. -3 15. -20 17. 14   19. -112 21. 73   23. -49
25. 20   27. -4

Objective C
29. 95   31. -960 33. -204 35. $799 profit

2.6 Quotient Rule and Integer Exponents

Getting Ready
1. -3 2. 3   3. 4096   4. $64x^3$ 5. 64   6. -125

Key Terms
1. zero   2. one   3. exponent   4. base   5. reciprocals

Objectives A, B, C
1. 3^3 3. x^3 5. z 7. 1   9. 1   11. 13   13. 9
15. $\dfrac{1}{y^7}$ 17. $\dfrac{1}{4^3 a^3}$ 19. $-\dfrac{8}{d^3}$ 21. $\dfrac{1}{a^{15}}$ 23. z^{12} 25. z^{24} 27. $\dfrac{1}{4^9}$
29. $\dfrac{1}{q^2}$

Answers to Worksheets for Classroom or Lab Practice

2.7 Power Rule for Quotients and Using Combined Laws of Exponents

Getting Ready
1. -243
2. 6
3. -24
4. $27x^6 y^3$
5. $a^8 b^{15}$
6. 1
7. $\dfrac{1}{c^4}$

Key Terms
1. factors
2. $\dfrac{a^n}{b^n}$
3. quotient
4. a^{m+n}
5. a^n
6. $a^n b^n$
7. invert
8. $\dfrac{1}{a^n}$
9. a^{mn}
10. $\left(\dfrac{b}{a}\right)^n$
11. a^{m-n}

Objective A
1. $\dfrac{x^2}{y^2}$
3. $\dfrac{625}{d^4}$
5. $\dfrac{36}{25}$

Objective B
7. $\dfrac{n^4}{m^4}$
9. $\dfrac{p^3}{8}$
11. $\dfrac{y^{12}}{x^{12}}$

Objective C
13. $\dfrac{81r^4}{s^4}$
15. $\dfrac{c^{32}}{d^{88}}$
17. $\dfrac{w^{20}}{z^{28}}$
19. $-\dfrac{x^{15}}{216}$
21. x^{22}
23. $\dfrac{1}{25p^6}$
25. $\dfrac{35}{m^6}$
27. $-\dfrac{z^{15}}{729}$
29. $\dfrac{16}{v^{20} w^{12}}$

2.8 Division of Polynomials by Monomials

Getting Ready
1. 6
2. $\dfrac{4}{x^4}$
3. $-5a^2$
4. 2
5. $216x^9 y^{15}$

Key Terms
1. term
2. monomial

Objective A
1. $7y^3$
3. $4d^3$
5. $c^6 d^4$
7. $-\dfrac{29 y^3}{x^5}$
9. $32a^6 b^{16}$
11. $\dfrac{2w^3}{7v}$

Answers to Worksheets for Classroom or Lab Practice

Objective B

13. $2x+6$
15. $2m^5-1$
17. $2c^3d^3-4c^5d^5$
19. $2z^2-\dfrac{8}{3z^2}$

21. $4y^2-\dfrac{20y^5}{x}+2x^6y^6$
23. $m^3+m^5+m^7$
25. $8v^6+7u^5v^8-4u$

27. $2x^4-4x^3+3x$
29. $\dfrac{3y^4}{z^2}$

2.9 An Application of Exponents: Scientific Notation

Getting Ready

1. 8
2. 14.35
3. 3.46
4. 10^5
5. 10^{-14}
6. 10^5

Key Terms

1. scientific notation
2. positive
3. coefficient
4. negative
5. exponential part

Objective A

1. 1. 260,000
3. .00032
5. 90,000,000

Objective B

7. 6.2×10^4
9. 2.6×10^{-6}
11. 9.37×10^5
13. 2.675×10^{10}

Objective C

15. 2.76×10^{10}; 27,600,000,000
17. 7.875×10^{-11}; .00000000007875
19. 2.0×10^4; 20,000
21. 6.0×10^8; 600,000,000
23. 3.45×10^8; 345,000,000
25. 5.25×10^{-1}; .525
27. 7.2×10^{-3}; .0072
29. 1.5×10^4; 15,000

Objective D

31. 1111.11 hr
33. 1.3×10^{21} times greater

Chapter 3 LINEAR EQUATIONS AND INEQUALITIES

3.1 Addition Property of Equality

Getting Ready

1. -2
2. -12
3. 1.2
4. -10.9
5. $7y+28$
6. $-5z+20$
7. $-2a+12$
8. $14y-16$
9. $3.5x+13.3$

Key Terms

1. equation
2. solution
3. linear
4. Addition Property of Equality

Answers to Worksheets for Classroom or Lab Practice

5. equivalent

Objective A
1. $m = -5$
3. $b = 9.8$
5. $x = -12$
7. $x = -6$
9. $x = 2$
11. $x = 15$
13. $x = 5$
15. $x = 0$

Objective B
17. $x - 8 = 17; x = 25$
19. $14 + x = 32; x = 18$
21. $x - 5 = -8; x = -3$

Objective C
23. 6 mi
25. $55
27. 178 lb

3.2 Multiplication Property of Equality

Getting Ready
1. -30
2. 48
3. -8
4. 16
5. $-.92$
6. -5
7. 52
8. 1
9. 1

Key Terms
1. reciprocal
2. product
3. Multiplication Property of Equality
4. one

Objective A
1. $x = -2$
3. $b = -19$
5. $n = 45$
7. $m = 32$
9. $c = -\dfrac{32}{15}$
11. $a = 6$

Objective B
13. $y = 5$
15. $m = 7$
17. $x = 4$
19. $w = -8$

Objective C
21. five times a number is thirty-five; the product of five and a number is thirty-five (answers vary)
23. negative seventeen is three times a number; negative seventeen is the product of three and a number (answers vary)
25. $7x = 42; x = 6$
27. $\dfrac{x}{-6} = -30; x = 180$
29. $\dfrac{x}{9} = 63; x = 567$
31. $8.25
33. 24 pieces

3.3 Combining Properties in Solving Linear Equations

Getting Ready
1. -25
2. 26
3. $-2a - 17$
4. $2x + 12$
5. -3.4
6. 7.2
7. -16
8. -19
9. $8x - 15$
10. $16x + 16$
11. $y = 16$
12. $a = 8$
13. $-.7$
14. -48

Answers to Worksheets for Classroom or Lab Practice

Key Terms
1. identity 2. contradiction

Objective A
1. $x = -2$ 3. $w = 4$ 5. $w = -\dfrac{15}{2}$ 7. $m = -\dfrac{4}{5}$ 9. $y = 5$
11. $x = -1$ 13. $b = -3$ 15. $n = -13$ 17. $x = -59$

Objective B
19. Identity, all real numbers   21. Contradiction, no solutions

Objective C
23. $3x - 6 = 15;\ x = 7$ 25. $4x - 7 = 29;\ x = 9$

Objective D
27. 60 in. by 72 in.   29. 5.4 hr

3.4 Using and Solving Formulas

Getting Ready
1. $a = 3$ 2. $y = 15$ 3. $x = 2.6$

Key Terms
1. variable   2. formula

Objective A
1. $A = 25$ in.^{2} 3. $r = 12$ ft   5. $P = 3200$ 7. $I = 4.8$
9. $T = 4.5$

Objective B
11. $h = \dfrac{2A}{b+B}$ 13. $I = \dfrac{P}{V}$ 15. $h = \dfrac{2A}{b}$ 17. $y = \dfrac{7-6x}{8}$
19. $t = \dfrac{A-P}{Pr}$

3.5 General, Consecutive Integer, and Distance Application Problems

Getting Ready
1. $x = 64{,}157$ 2. $x = 520$ 3. $x = 17$ 4. $x = -\dfrac{10}{13}$

Key Terms
1. consecutive integers   2. consecutive even integers
3. consecutive odd integers   4. time

Answers to Worksheets for Classroom or Lab Practice

Objective A
1. 652 boys
3. 5280 votes
5. 32

Objective B
7. 1235, 1236
9. 122, 123, 124
11. 18
13. 31, 32

Objective C
15. 104 mi
17. 12 noon
19. 300 mi

3.6 Money, Investment, and Mixture Application Problems

Getting Ready
1. $x = 4$
2. $x = 3$
3. $x = 1.7$

Key Terms
1. investment problems
2. mixture problems
3. $500 - x$
4. $500 + x$

Objective A
1. 21 inside doors, 7 outside doors
3. 20 packs of pencils, 18 packs of pens
5. 45 bags

Objective B
7. $900 at 6%, $1700 at 7.5%
9. $2400

Objective C
11. 15 gal
13. 32 lb of dried fruit, 8 lb of nuts

3.7 Geometric Application Problems

Getting Ready
1. $x = 6$
2. $y = -3$
3. $z = 9$

Key Terms
1. supplementary
2. complementary
3. degrees
4. triangle
5. parallel
6. transversal
7. congruent

Objective A
1. 50 ft
3. 6 ft, 24 ft, 31 ft
5. 45°, 45°, 90°
7. 36°, 144°
9. 500 ft, 600 ft, 700 ft
11. 138°
13. 16°, 74°
15. $x = 18$, $m\angle A = 63°$, $m\angle B = 63°$
17. $x = 16$, $m\angle A = 134°$, $m\angle B = 46°$

3.8 Solving Linear Inequalities

Getting Ready
1. $a = 15$
2. $b = 7$
3. $x = -4$
4. $y = -32$

Answers to Worksheets for Classroom or Lab Practice

5. $x = -\dfrac{3}{4}$ 6. $x = -\dfrac{20}{3}$

Key Terms
1. direction 2. irrational 3. inequality 4. dot
5. open dot 6. compound 7. Addition Property of Inequality

Objective A
1.
3.
5.
7.

Objective B
9. $x \geq 8$
11. $q > 2$
13. $n < -7$

Objective C
15. $y \geq 2$ 17. $w > -12$ 19. $q < .2$

Objective D
21. $z \geq -3$ 23. $b < -9$ 25. $y \leq -9$

Objective E
27. $-5 < x \leq 4$ 29. $-\dfrac{5}{6} \leq y \leq \dfrac{7}{12}$

Objective F
31. $3x > -24$; $x > -8$ 33. $9(x-1) \geq 36$; $x \geq 5$

Chapter 4 GRAPHING LINEAR EQUATIONS AND INEQUALITIES

4.1 Reading Graphs and the Cartesian Coordinate System

Getting Ready
1. 2 2. 15 3. −40 4. −6 5. −26
6. 34 7. $x = 2$ 8. $x = -8$

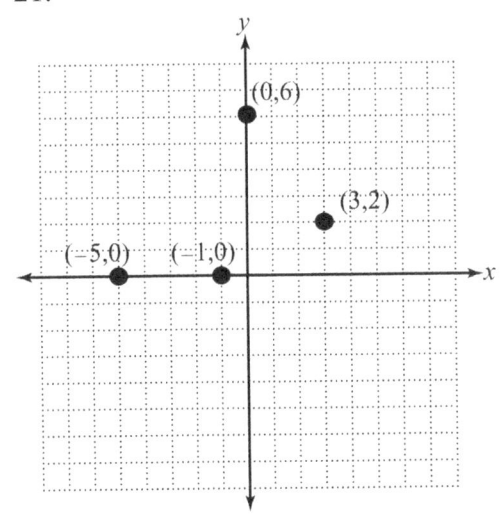

Answers to Worksheets for Classroom or Lab Practice

Key Terms
1. circle graph
2. quadrants
3. x-axis
4. y-axis
5. origin
6. ordered pair
7. scatter diagram

Objective A
1. Math
3. 159 new hires
5. Niger, Zambia, Haiti
7. 2005-2006
9. $174,000 million

Objective B, C
11. $6 - 3(-1) = 9$
    $6 + 3 = 9$
    $9 = 9$ ✓

    $(6, -1)$

Objective D
13. not a solution
15. solution

Objective E
17. $(0,7), \left(\dfrac{7}{3}, 0\right), (2,1)$
19. $(0,4), (-5,0), (-10,-4)$

Objective F
21.

Objective G
23. (January, 377), (February, 387), (March, 552), (April, 467), (May, 484), (June, 515), (July, 425), (August, 448), (September, 383), (October, 372), (November, 362), (December, 472)

Answers to Worksheets for Classroom or Lab Practice

25.

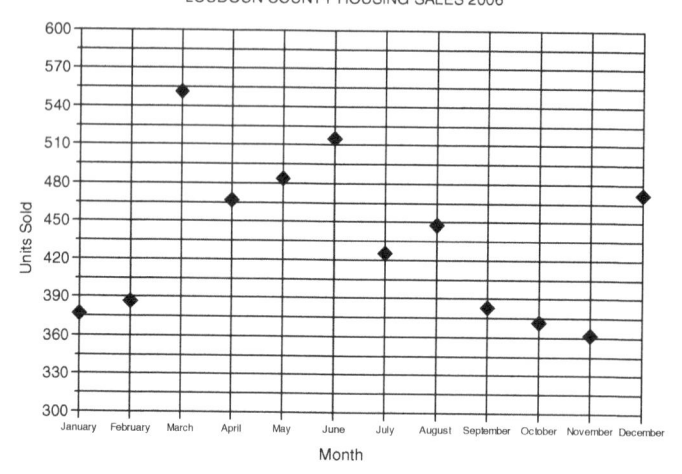

[Graph: LOUDOUN COUNTY HOUSING SALES 2006 — Units Sold vs. Month; data points approximately: January 375, February 385, March 550, April 465, May 485, June 515, July 425, August 445, September 380, October 365, November 360, December 470. (Source: Northern Virginia Association of Realtors®)]

4.2 Graphing Linear Equations with Two Variables

Getting Ready
1. -11
2. -8
3. 24
4. 3
5. 16
6. 53
7. $x = 5$
8. $x = -3$
9. [Graph showing points $(-4, 1)$, $(3, -2)$, and $(0, -4)$ plotted on the xy-plane.]

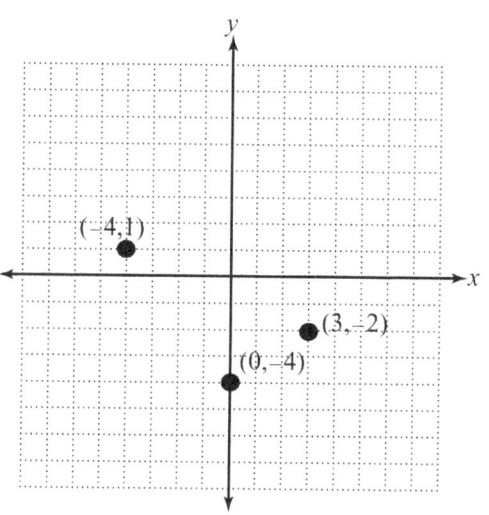

Key Terms
1. straight line
2. linear equation
3. vertical
4. graph
5. horizontal

Answers to Worksheets for Classroom or Lab Practice

Objective A

1. $(0,1), (-1,-1), (1,3)$

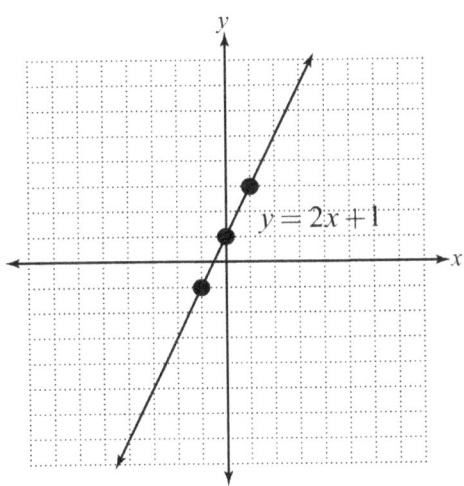

3. $(0,5), \left(\frac{5}{3},0\right), (1,2)$

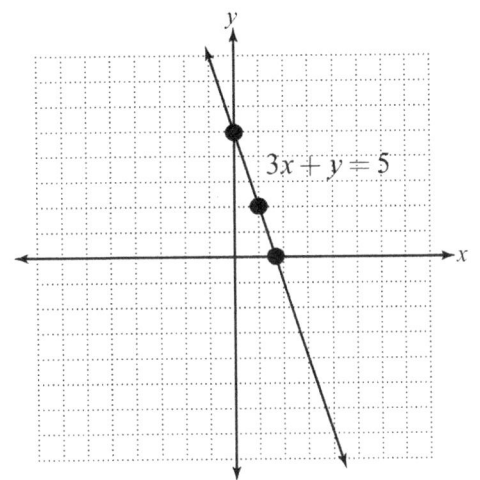

5. $(0,6), (-1,7), (6,0)$

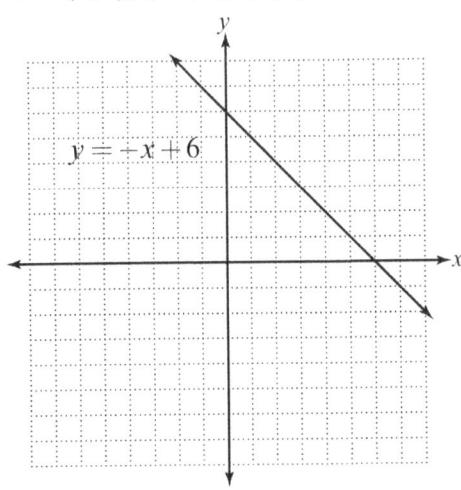

7. $(0,-2), (1,1), (-1,-5)$

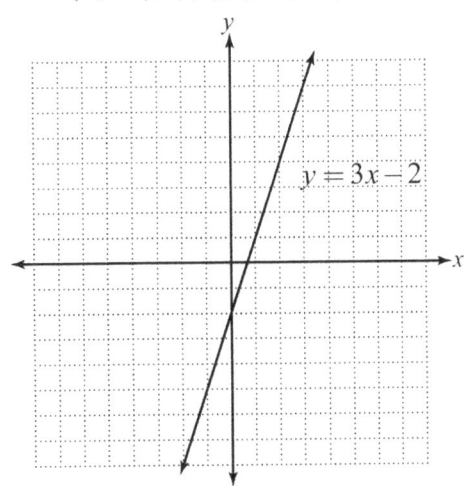

9. $(0,-8), (8,0), (1,-7)$

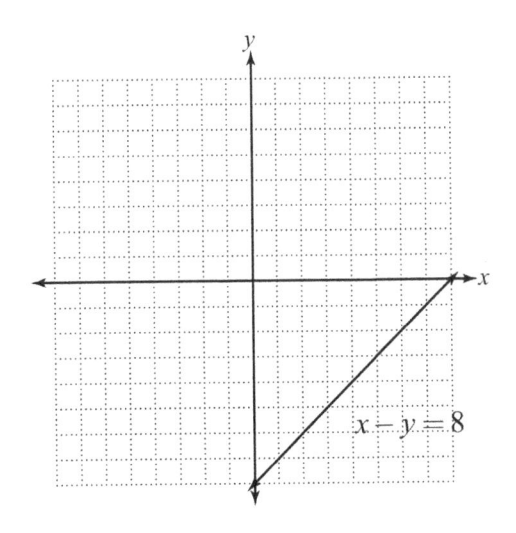

11. $(0,-5), (7,0), \left(1,-\frac{30}{7}\right)$

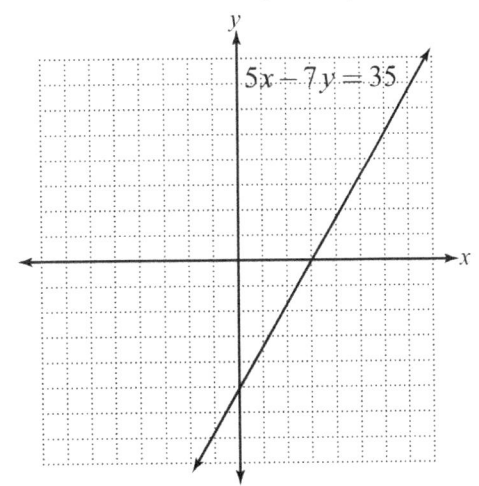

Copyright © 2013 Pearson Education, Inc.   **461**

Answers to Worksheets for Classroom or Lab Practice

13. $(0,-3)$, $(2,4)$, $\left(\frac{6}{7},0\right)$

$7x-2y=6$

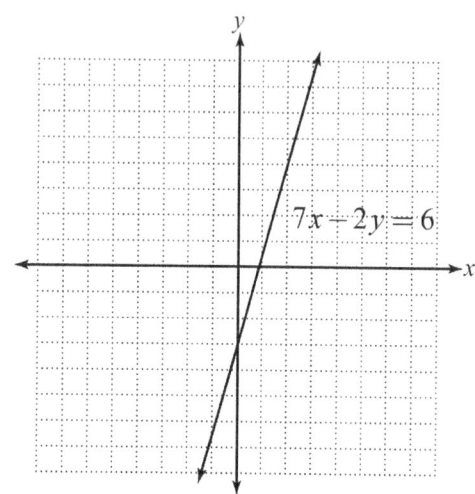

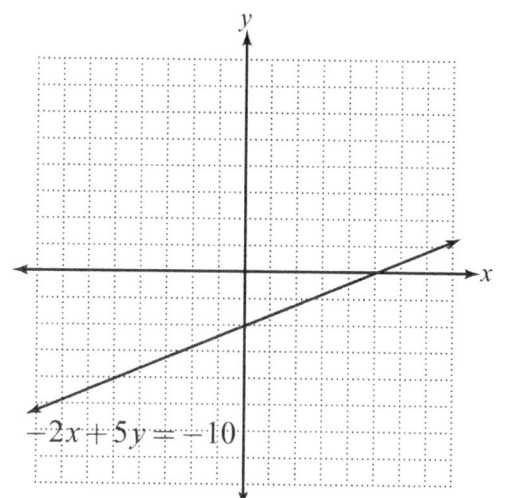

15. $(0,-2)$, $(5,0)$, $(-5,-4)$

$-2x+5y=-10$

17. $(0,0)$, $(1,7)$, $(-1,-7)$

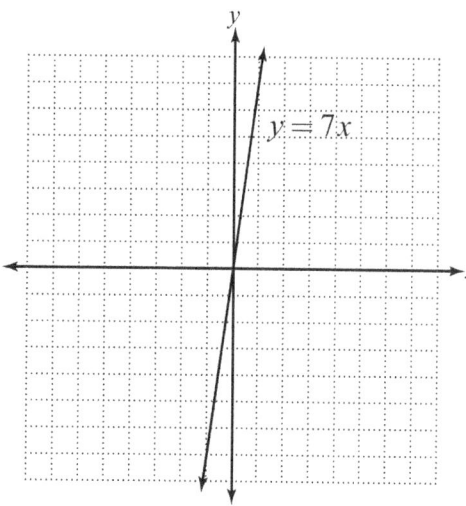

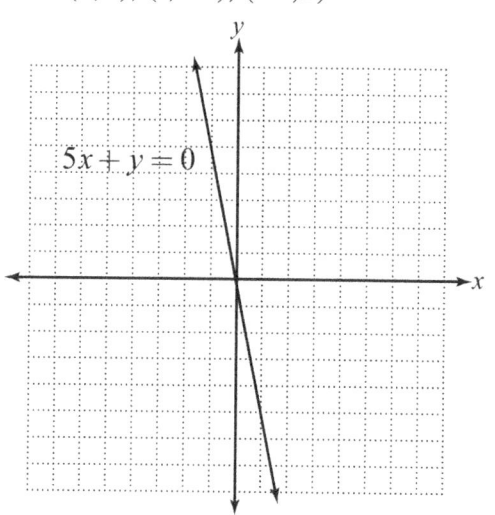

$y=7x$

19. $(0,0)$, $(1,-5)$, $(-1,5)$

$5x+y=0$

21. $(0,0)$, $(4,-1)$, $(-4,1)$

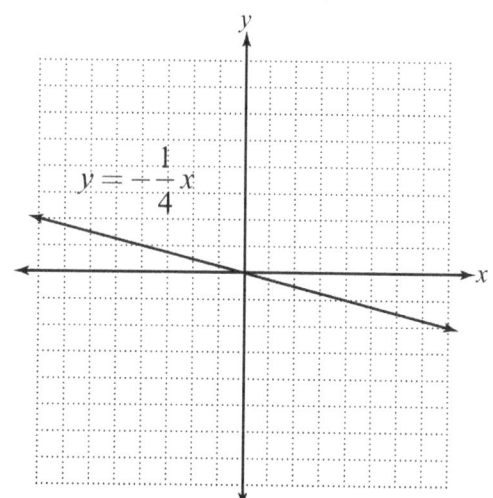

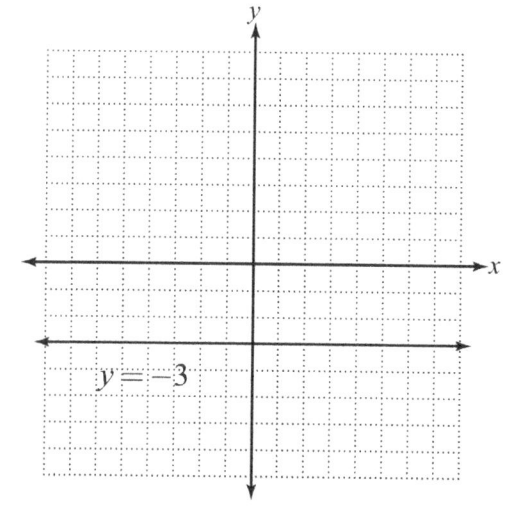

$y=-\frac{1}{4}x$

23. $(0,-3)$, $(-1,-3)$, $(1,-3)$

$y=-3$

462　　　　　Copyright © 2013 Pearson Education, Inc.

Answers to Worksheets for Classroom or Lab Practice

25. $(0,5), (-1,5), (1,5)$ 27. 70 miles   29. 95 miles

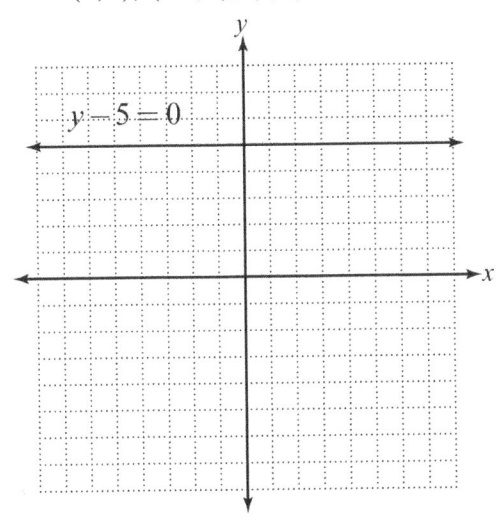

4.3 Graphing Linear Equations by Using Intercepts

Getting Ready
1. -1 2. 21 3. -12 4. -4 5. -15
6. 60 7. $x=1$ 8. $y=-4$
9.   10. $A(5,3), B(-5,0), C(0,-2), D(5,-4)$

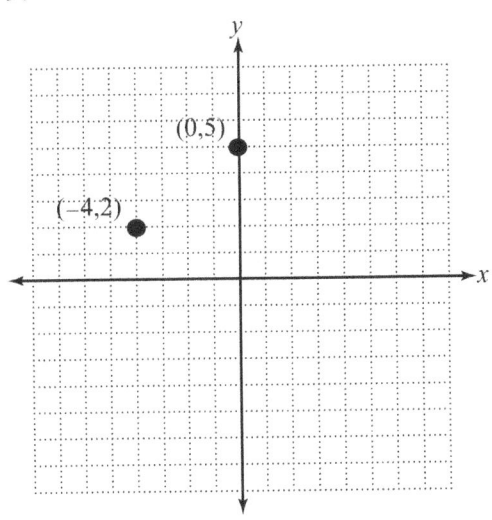

Key Terms
1. x-intercept   2. y-intercept   3. y 4. x 5. $x=k$
6. $y=k$

Objective A
1. $(6,0), (0,-6)$ 3. $(3,0), (0,6)$ 5. $(4,0)$, no y-intercept

Answers to Worksheets for Classroom or Lab Practice

Objective B

7.

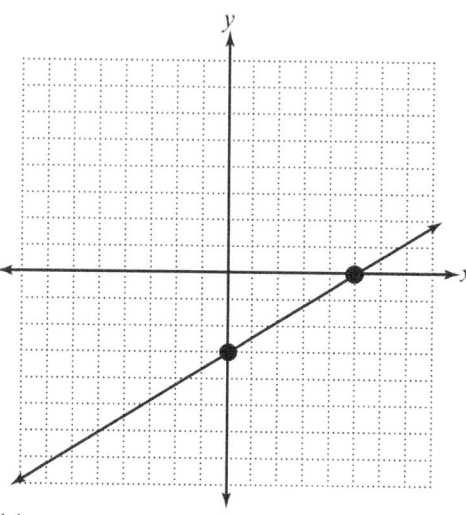

9.

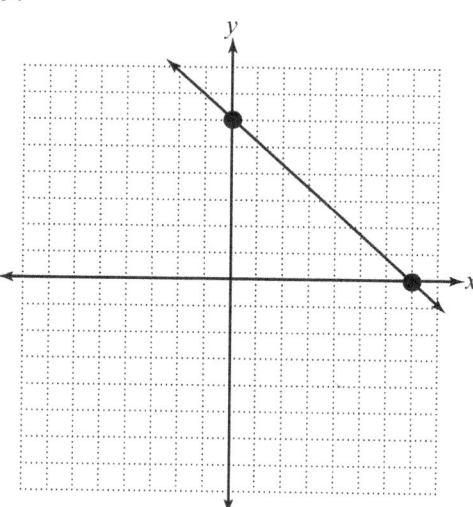

11. $x+y=7$

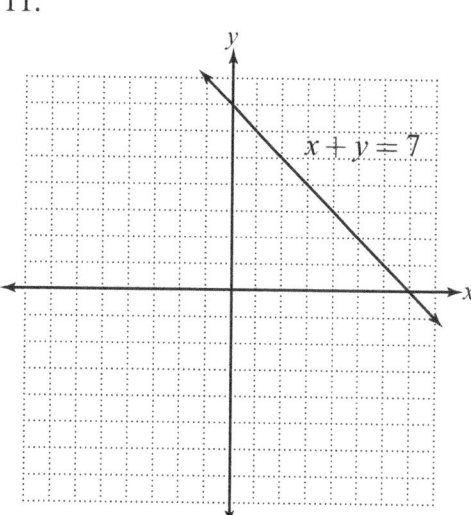

13. $4x-y=8$

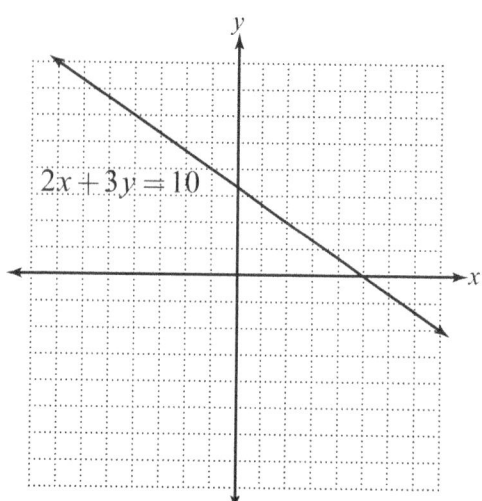

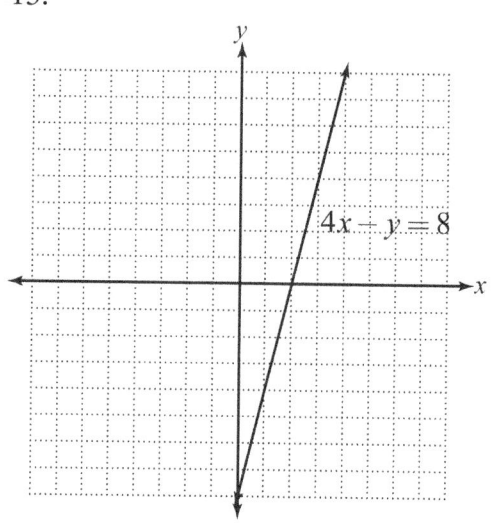

15. $2x+3y=10$

Answers to Worksheets for Classroom or Lab Practice

Objective C

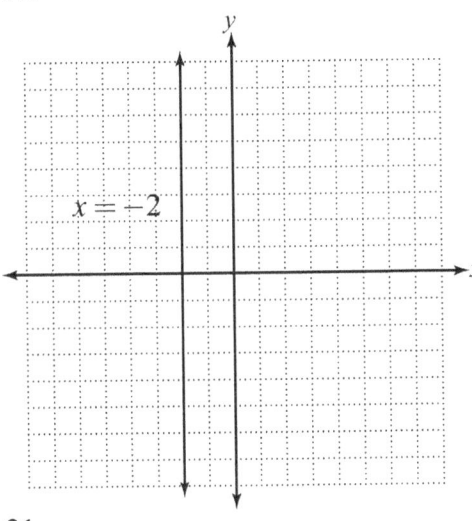

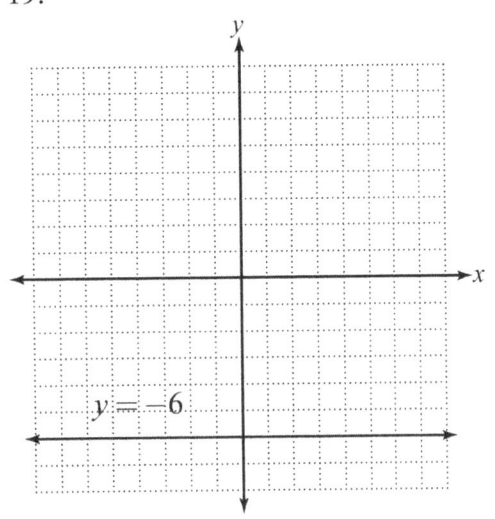

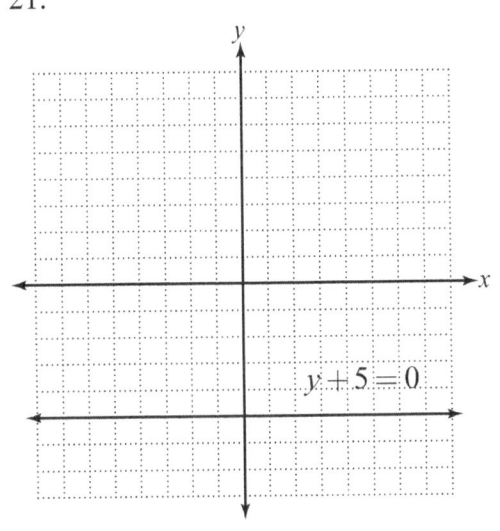

17. [Graph showing vertical line $x = -2$]

19. [Graph showing horizontal line $y = -6$]

21. [Graph showing horizontal line $y + 5 = 0$]

4.4 Slope of a Line

Getting Ready

1. -12 2. 11   3. 54   4. 3   5. 6   6. 3

Copyright © 2013 Pearson Education, Inc.   465

Answers to Worksheets for Classroom or Lab Practice

7.

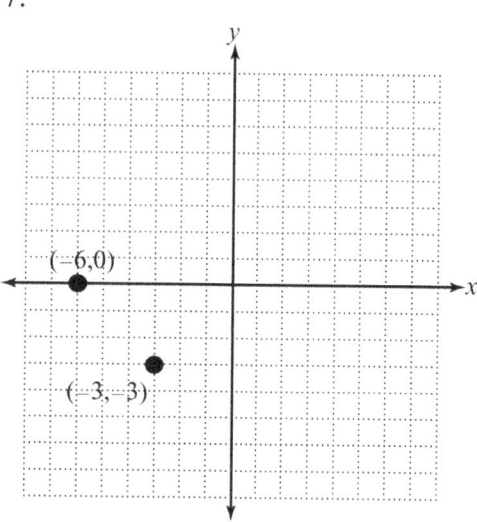

Key Terms

1. falls   2. slope   3. rises   4. undefined   5. zero

Objective A

1. $\dfrac{1}{8}$ 3. $\dfrac{1}{3}$ 5. $\dfrac{8}{7}$ 7. -4

Objective B

9. 3   11. $\dfrac{1}{2}$ 13. 0

Objective C

15.

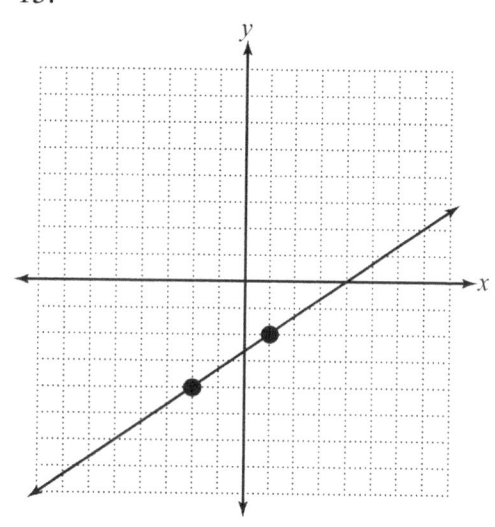

17.

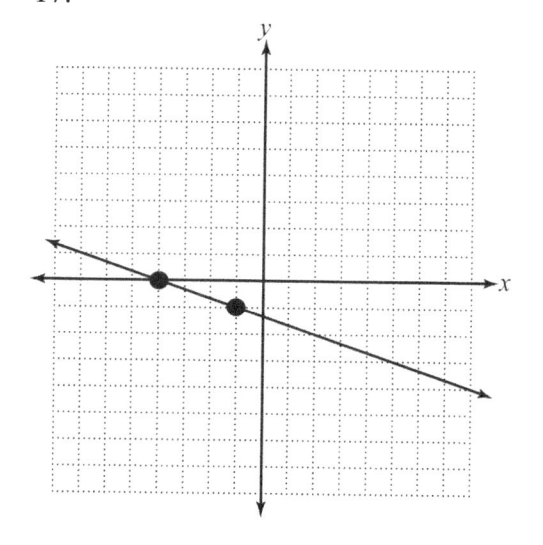

Answers to Worksheets for Classroom or Lab Practice

19.

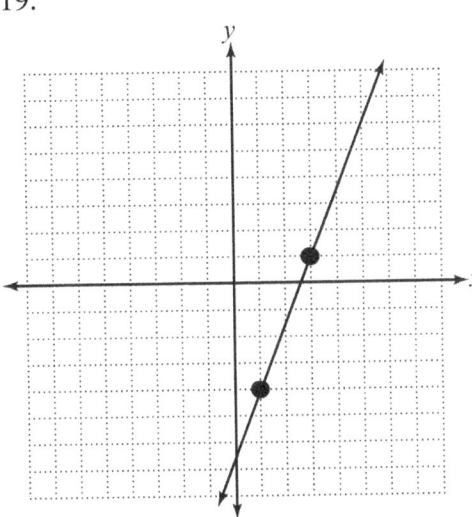

21. undefined   23. 0

4.5 Slope-Intercept Form of a Line

Getting Ready

1. -13 2. 12 3. 52 4. -3 5. $-\dfrac{1}{3}$

6.

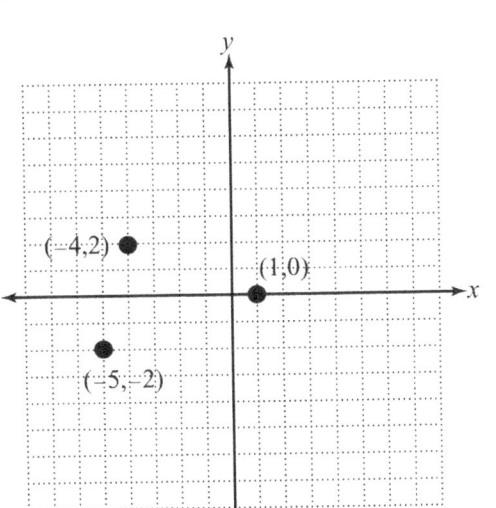

7. $y = -5x + 7$ 8. $y = \dfrac{6}{7}x - 2$

Key Terms
1. perpendicular 2. parallel 3. b 4. m
5. slope-intercept

Answers to Worksheets for Classroom or Lab Practice

Objective A, B

1. $m = 3$; y-intercept: 4

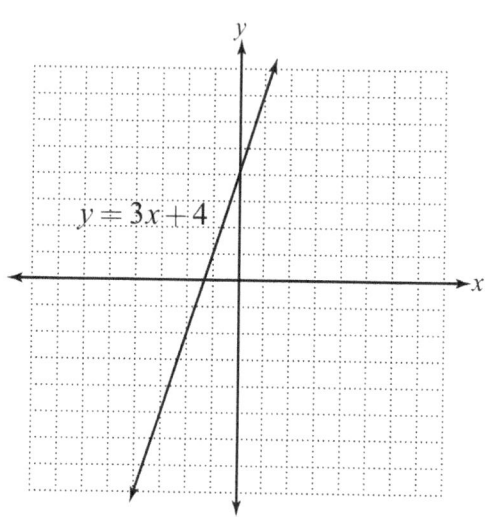

3. $m = -1$; y-intercept: 5

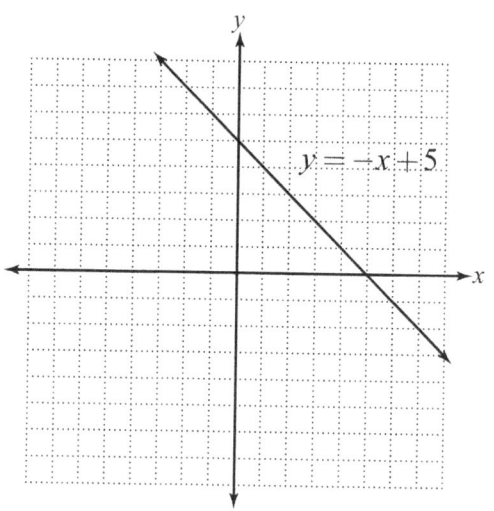

5. $m = -\dfrac{2}{7}$; y-intercept: -1

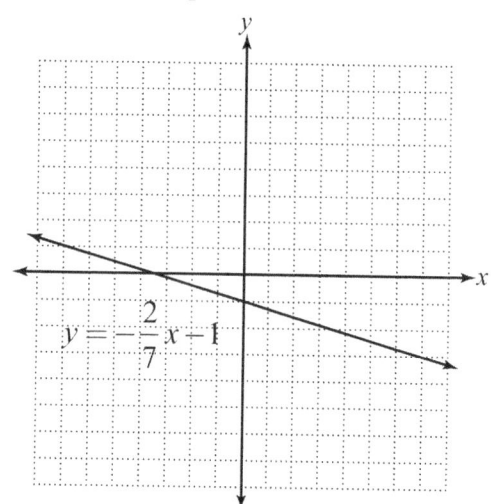

7. $m = -\dfrac{2}{5}$; y-intercept: 3

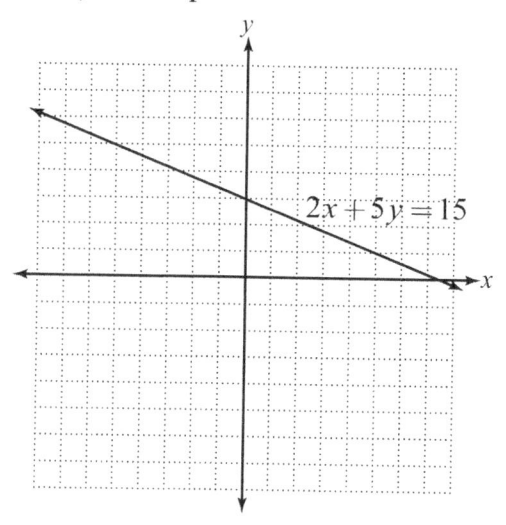

Answers to Worksheets for Classroom or Lab Practice

9. $m = \dfrac{3}{4}$;

 y-intercept: $-\dfrac{9}{4}$

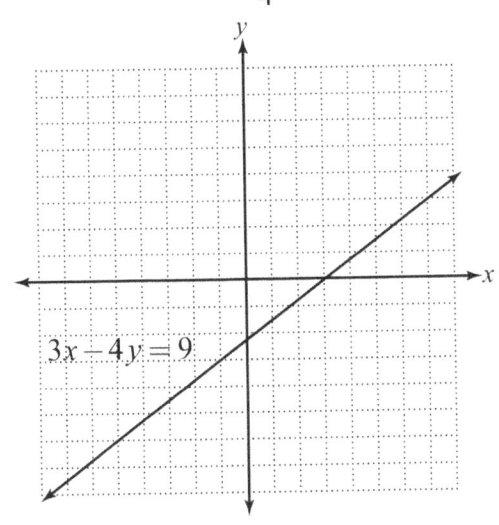

[Graph showing line $3x - 4y = 9$]

Objective C

11. $y = 3x - 5$ 13. $y = -\dfrac{2}{3}x - 8$

Objective D

15. $x = -2$

Objective E

17. parallel 19. perpendicular 21. neither 23. perpendicular
25. parallel

4.6 Point-Slope Form of a Line

Getting Ready

1. 22 2. -7 3. 15 4. -4 5. $7x - 14$
6. $x - 3y = -7$ 7. $y = \dfrac{1}{3}x + \dfrac{7}{3}$ 8. 1 9. $\dfrac{4}{5}$
10. neither 11. perpendicular 12. 2200

Key Terms

1. equal 2. vertical 3. horizontal 4. negative reciprocals
5. point-slope formula

Objective A

1. $6x - y = 37$ 3. $x - 5y = 49$ 5. $3x - 5y = -63$ 7. $x = -9$
9. $x = 2$

Copyright © 2013 Pearson Education, Inc. 469

Answers to Worksheets for Classroom or Lab Practice

Objective B, C

11. $y = -\dfrac{5}{3}x - \dfrac{11}{3}$ 13. $y = \dfrac{2}{9}x - \dfrac{8}{9}$

Objective D

15. $7x - 3y = 1$ 17. $3x + 8y = 23$ 19. $y = 14$ 21. $x = 0$

23a. $y = 240x$ b. $1920

4.7 Graphing Linear Inequalities with Two Variables

Getting Ready

1. -12 2. 12 3. 66 4. 6 5. 39 6. -40

7. [graph of $6x + 4y = 20$]

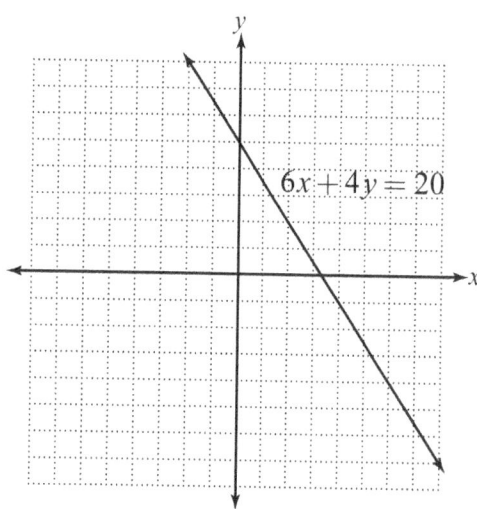

8. [graph of $3x - y = 0$]

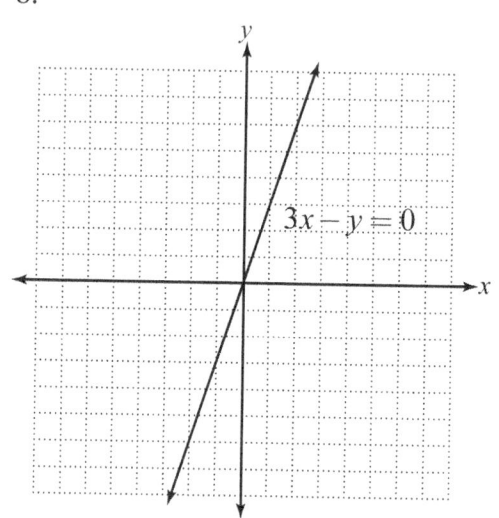

9. [graph of $y = -5$]

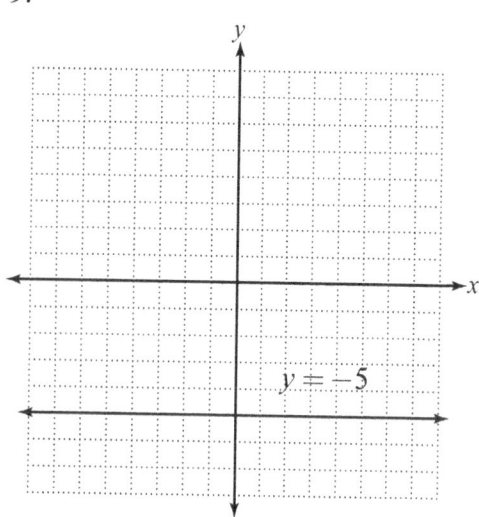

10. [graph of $x = 4$]

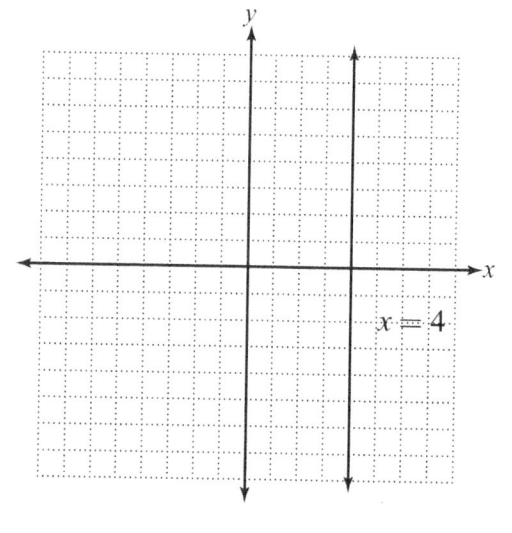

Key Terms

1. linear inequality   2. solid   3. dashed

Answers to Worksheets for Classroom or Lab Practice

Objective A
1. not a solution
3. not a solution
5. $y \leq x - 5$
7. $y \geq \frac{1}{4}x + 2$
9. $x + 5y > 5$
11. $y < 4x$
13. $x \geq 3$
15. $x < -2$

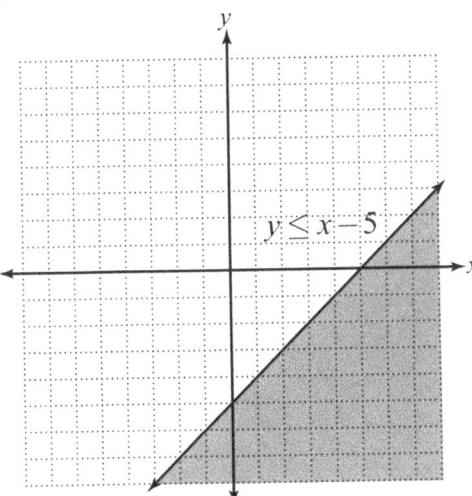

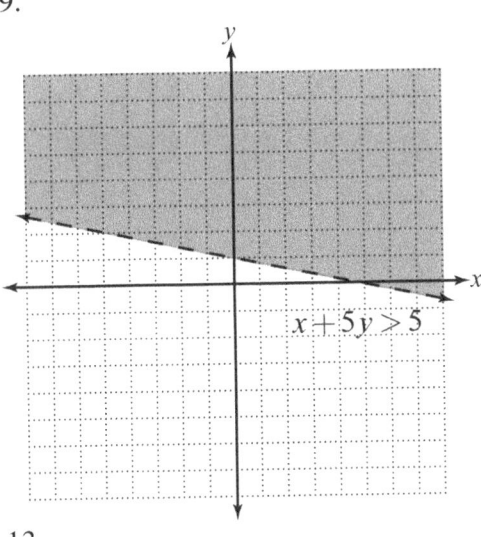

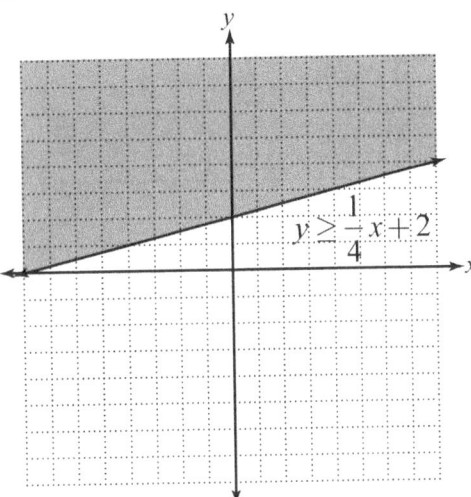

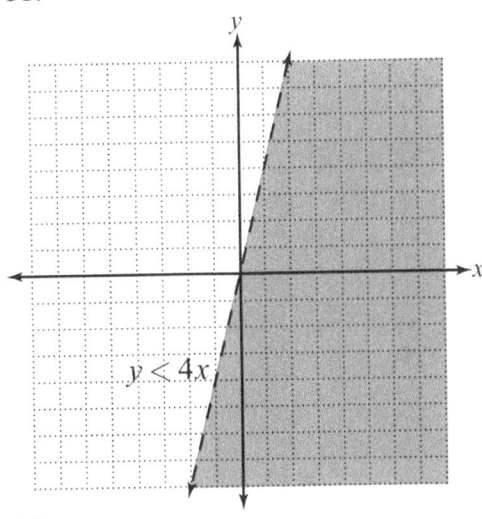

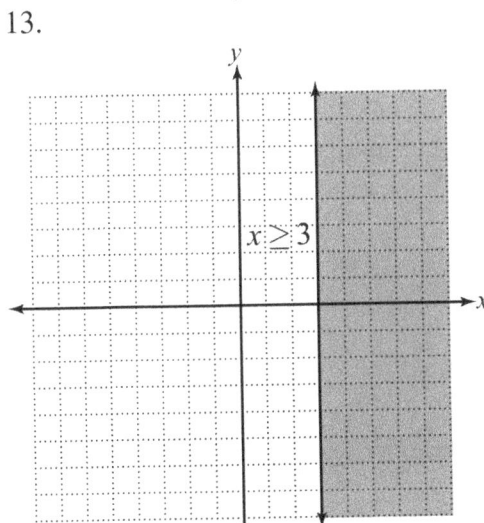

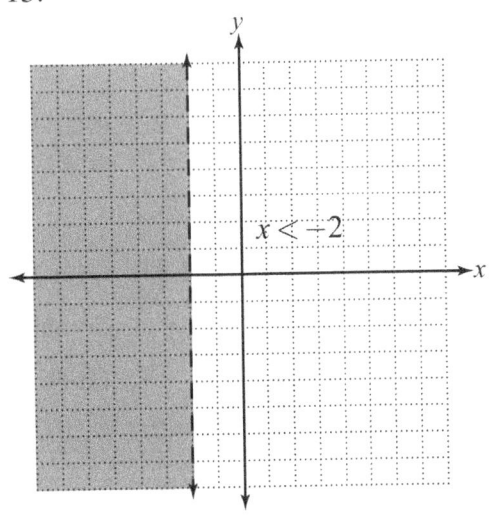

Answers to Worksheets for Classroom or Lab Practice

17.

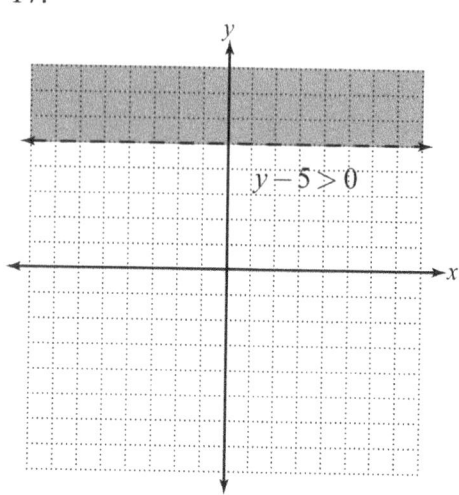

[Graph showing shaded region above horizontal line, labeled $y - 5 > 0$]

4.8 Relations and Functions

Getting Ready

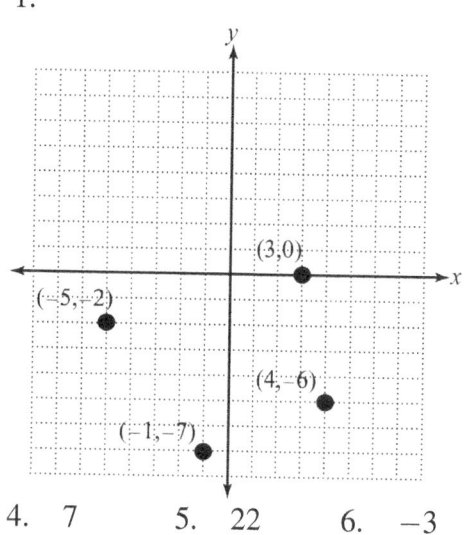

1. [Graph with points (3,0), (-5,-2), (4,-6), (-1,-7)]

2. neither   3. vertical

4. 7   5. 22   6. −3   7. −54

Key Terms

1. elements   2. relation   3. domain   4. range
5. function   6. vertical line   7. evaluating

Objective A

1. function;
   domain: {−3, 0, 1}
   range: {4, 5, 6}

3. not a function;
   domain: {−3, 3, 5}
   range: {2, 5, 6, 8}

Objective B

5. not a function   7. function   9. not a function

11. function   13. function

Objective C
15. (−1, 10)   17. (3, −6)   19. (4, 19)   21. (0, 3)
23. $1600

Chapter 5   FACTORS, DIVISORS, AND FACTORING

5.1   Prime Factorization and Greatest Common Factor

Getting Ready
1. 27   2. 63   3. $2^2 \cdot 3^2 \cdot 4^3$ 4. $3 \cdot 5^3 \cdot 7^2$ 5. 48   6. 60

Key Terms
1. composite factor   2. divisible   3. prime   4. greatest common
5. factorization   6. prime factorization

Objective A
1. prime   3. prime   5. 1, 2, 4, 17, 34, 68   7. 1, 2, 11, 22

Objective B
9. 2   11. 3   13. none of these

Objective C
15. $3^2 \cdot 5^2 \cdot 7$ 17. $7 \cdot 11 \cdot 17$

Objective D
19. 5   21. 8   23. 7

Objective E
25. ab^3 27. $6w^2 z^3$ 29. $2pq$

5.2   Factoring Polynomials with Common Factors and by Grouping

Getting Ready
1. $9a + 12b$ 2. $-24x - 20y$ 3. $2y^5 - 6y^2 z$ 4. $b^4 y^2 + b^3 y^3$
5. $3a^3 b - 5a^2 b^2 - 6ab^3$ 6. $3x^3$ 7. $2b^3$ 8. -24 9. 40

Key Terms
1. distributive property   2. factorization   3. factoring by grouping
4. greatest common factor

Objective A
1. $w(m - n)$ 3. $x^3(x^2 + 1)$ 5. $16a^3(2a - 1)$ 7. $12z^3(4 - z^3)$

Answers to Worksheets for Classroom or Lab Practice

9. $pq(q^3+p)$
11. $2(x^2-3x+9)$
13. $2(3m^2+2m+4)$
15. $7xy^3(x+3y^2-5x^2y)$
17. $-11(4z-3)$
19. $(c+d)(a-2)$
21. $(y+8)(x+z)$

Objective B
23. $(x+2)(y+4)$
25. $(3w-z)(7w+z)$
27. $(b+3d)(2a+c)$
29. $(5x-y)(3x+2)$

5.3 Factoring General Trinomials with Leading Coefficients of One

Getting Ready
1. y^2+y-12
2. $x^2-13x+42$
3. x^4y^3
4. $3a$
5. $6ab$
6. -48
7. 36
8. $3(x-2)(x+1)$
9. $5n^2(m-5)(m-3)$

Key Terms
1. positive
2. negative
3. prime
4. trinomial

Objective A
1. $(z-12)(z+8)$
3. $(b+6)(b-1)$
5. prime
7. $(c-11)(c+8)$
9. $(x+20)(x+2)$
11. prime
13. $(m+9n)(m+2n)$
15. $(x+2y)(x-y)$
17. $(c-7d)(c+d)$
19. $(r+5t)(r-4t)$
21. $5(x+2)(x-1)$
23. $4(a+6)(a+5)$
25. $10(y-6)(y-7)$
27. $a(a+9)(a+4)$
29. $p^2(p-7)(p+1)$

5.4 Factoring General Trinomials with Leading Coefficients Other Than One

Getting Ready
1. $(x-7)(x+9)$
2. $(x+4)(x+1)$
3. $w^3(w-3)$
4. $bc^2(c^3+c-5)$

Key Terms
1. leading
2. greatest common factor
3. ac

Objective A
1. $(2a+5)(a-3)$
3. prime
5. $(5b+4)(b+3)$
7. $(2p+5)(3p+1)$
9. $(3a-4)(2a-5)$
11. $(3m-1)(m+7)$
13. $2(2m+7)(9m+1)$
15. $(9p-2)(4p+3)$
17. $(11r-2)(3r-2)$
19. $(5b+3)(3b+5)$
21. $(10m+7)(2m+9)$
23. prime
25. $(18y+5)(y-3)$
27. $(3p+4q)(2p-7q)$
29. $(7m-6)(6m-7)$

Answers to Worksheets for Classroom or Lab Practice

5.5 Factoring Binomials

Getting Ready
1. $x^2 - 16$
2. $4x^2 - 25y^2$
3. a^9
4. $25x^4$
5. $5(2x^2 + 5y^2)$
6. $15x^3 + 4x^2 + 3x - 2$

Key Terms
1. conjugate pairs
2. perfect cubes
3. perfect squares
4. binomial
5. $(a-b)(a^2 + ab + b^2)$
6. $(a+b)(a^2 - ab + b^2)$

Objective A
1. $(p+q)(p-q)$
3. prime
5. $(7w+9z)(7w-9z)$
7. $(10m+1)(10m-1)$
9. $(8p+9)(8p-9)$
11. $(3z+25)(3z-25)$
13. $(5m^2 + n)(5m^2 - n)$
15. $(x^2 + 25)(x+5)(x-5)$
17. $(9x^2 + 4)(3x+2)(3x-2)$

Objective B
19. $(10-a-b)(10+a+b)$
21. $(2m+2n+7)(2m+2n-7)$
23. $7(x+2)(x-2)$

Objective C
25. $(d+f)(d^2 - df + f^2)$
27. $(x+5)(x^2 - 5x + 25)$
29. $(10m-n)(100m^2 + 10mn + n^2)$
31. $(7w-6)(49w^2 + 42w + 36)$

5.6 Factoring Perfect Square Trinomials

Getting Ready
1. $x^2 - 6xy + 9y^2$
2. $16a^2 + 24ab + 9b^2$
3. $60xy$
4. $-48c$
5. $ab(b^2 + 3b + 10)$

Key Terms
1. $(a-b)^2$
2. prime
3. perfect square trinomial

Objective A
1. perfect square trinomial
3. perfect square trinomial

Objective B
5. $(a+7)^2$
7. $(b-5)^2$
9. $(2t+3)^2$
11. $(4q+7)^2$
13. $(6a-5b)^2$
15. $3(n+5p)^2$
17. $m^2 n(m-n)^2$
19. $5xy(2x+3y)^2$

Answers to Worksheets for Classroom or Lab Practice

5.7 Mixed Factoring

Getting Ready
1. $a(a-4)(a-5)$
2. $(5x-12)(5x+12)$
3. $3(3c-1)(c+1)$
4. $(x+y)(a+b)$

Key Terms
1. greatest common factor
2. three
3. two
4. grouping
5. prime

Objective A
1. $3(8-7x^2)$
3. $(5+y)(5-y)$
5. prime
7. $2(y+7)(y-8)$
9. $3c(c-1)(c-2)$
11. $6p^2q(2p+1)(p-1)$
13. $16(n+5)(n-5)$
15. $(a+b)(2c-3d)$
17. $(5p-3q)(6a+b)$
19. $(n+p)(m-7)$
21. $(3m-2n)(6x+5y)$
23. prime
25. $(2a-b+2)(2a-b-2)$
27. $27(y+2)(y^2-2y+4)$
29. $n^2(n+2)(n^2-2n+4)$

5.8 Solving Quadratic Equations by Factoring

Getting Ready
1. $p(p-9)$
2. $(4x-3)(x-1)$
3. $2(5t+3)(t-2)$
4. $x=4$
5. $x=\dfrac{5}{2}$

Key Terms
1. quadratic
2. parabolas
3. zero product
4. zero

Objective A
1. 2, 11
3. $-\dfrac{7}{2}, -\dfrac{3}{5}$
5. 6, 9
7. 0, −10
9. $-\dfrac{1}{8}, -\dfrac{1}{9}$
11. 4, −4
13. 2, −6
15. $0, -4, -\dfrac{1}{3}$
17. −4, 5
19. $-\dfrac{4}{5}, \dfrac{1}{3}$

Objective B
21. 6 ft by 9 ft
23. 5 sec

Objective C
25. 8

Answers to Worksheets for Classroom or Lab Practice

Chapter 6 MULTIPLICATION AND DIVISION OF RATIONAL NUMBERS AND EXPRESSIONS

6.1 Reducing Rational Numbers and Rational Expressions

Getting Ready
1. $2 \cdot 3^3$
2. $2^2 \cdot 3 \cdot 13$
3. $x^3 \cdot y^2$
4. $a^5 \cdot b^4$

Key Terms
1. lowest terms
2. equivalent
3. rational expression
4. rational

Objective A

1. Number line with points at $-1\frac{1}{3}$, $-\frac{3}{4}$, $2\frac{1}{2}$ marked between -3 and 3.

Objective B
3. $\dfrac{1}{9}$
5. $\dfrac{3}{16}$
7. $\dfrac{8}{7}$
9. $\dfrac{3}{20}$
11. $\dfrac{81}{70}$
13. $\dfrac{5}{4}$

Objective C
15. $\dfrac{y^2}{x^3}$
17. $-\dfrac{p^6}{q^4}$
19. $-\dfrac{5z^5}{x^2 y^3}$
21. $\dfrac{a^5}{5b^2 c^7}$
23. $-\dfrac{x^2 y}{3}$
25. $\dfrac{2b}{3ac^3}$

6.2 Further Reduction of Rational Expressions

Getting Ready
1. undefined
2. $x = -\dfrac{5}{3}$
3. $x = -3, 7$
4. $\dfrac{7}{9}$
5. $2(x+3)$
6. $(x+1)(x+5)$
7. $(x+5)(x-5)$
8. 1

Key Terms
1. terms
2. factor
3. undefined
4. rational

Objective A
1. $x \neq 4$
3. $m \neq -1$
5. $p \neq -6, 7$
7. $x \neq -5, -3$

Objective B
9. $\dfrac{3}{7}$
11. $\dfrac{2}{9}$
13. $-\dfrac{7}{8}$
15. $\dfrac{x-1}{x+3}$
17. $\dfrac{p+3}{p-5}$
19. $\dfrac{3a+b}{6a-b}$
21. $-x-7$

Copyright © 2013 Pearson Education, Inc.   477

Answers to Worksheets for Classroom or Lab Practice

6.3 Multiplication of Rational Numbers and Expressions

Getting Ready
1. $2^3 \cdot 3$
2. $\dfrac{22}{5}$
3. $2^2 \cdot 3 \cdot x^3 \cdot y^5$
4. $30a^5b^5$

Key Terms
1. product
2. reduce
3. improper fractions

Objective A
1. $\dfrac{8}{63}$
3. $\dfrac{21}{50}$
5. $\dfrac{1}{4}$
7. $\dfrac{24}{5}$
9. $\dfrac{1}{8}$

Objective B
11. 22
13. $\dfrac{391}{75}$
15. $-\dfrac{70}{9}$
17. $-\dfrac{22}{7}$
19. -106

Objective C
21. $\dfrac{x^2}{y^5}$
23. $\dfrac{1}{p^4 q^4}$
25. $\dfrac{2cd^2}{3}$
27. $\dfrac{4c^5}{3a^5 b^2}$
29. 38 ft

6.4 Further Multiplication of Rational Expressions

Getting Ready
1. $\dfrac{3}{8}$
2. $\dfrac{1}{y^2}$
3. $6(x-3)$
4. $(x+7)(x-5)$
5. $(x+9)(x-9)$

Key Terms
1. factor
2. divide
3. -1

Objective A
1. $\dfrac{3w(x+y)^2}{2}$
3. $\dfrac{1}{4}$
5. $\dfrac{x+4}{x+3}$
7. $\dfrac{6a+1}{(6a-1)(2a-5)}$
9. $\dfrac{x+3}{x+5}$
11. $\dfrac{3x-4}{3x+4}$

Objective B
13. -2
15. $-(x+2)(x+3)$
17. $-\dfrac{4x+5}{x-8}$
19. $-\dfrac{y-6}{y-3}$

Answers to Worksheets for Classroom or Lab Practice

6.5 Division of Rational Numbers and Expressions

Getting Ready
1. $\dfrac{20}{3}$
2. $2 \cdot 3^2$
3. $x(x-8)$
4. $(2x+1)(x-5)$
5. $(x+5)(x-5)$
6. $\dfrac{1}{20}$
7. a
8. $\dfrac{x}{x+5}$

Key Terms
1. reciprocal
2. multiplicative inverses
3. divide
4. expressions

Objective A
1. $\dfrac{2}{3}$
3. $\dfrac{4}{5}$
5. $-\dfrac{48}{7}$
7. $-\dfrac{70}{11}$
9. $\dfrac{105}{53}$
11. $\dfrac{38}{27}$

Objective B
13. $12

Objective C
15. xy^3
17. $\dfrac{2q^4 y^2}{9px^3}$
19. $\dfrac{-2(x-4)}{y^2}$
21. $-\dfrac{p+9}{p+2}$
23. $\dfrac{y-5}{y+5}$
25. $-\dfrac{a+2}{a-3}$

6.6 Division of Polynomials (Long Division)

Getting Ready
1. 24
2. $15a^5$
3. $3a^3$
4. $-x$
5. $4x$

Key Terms
1. quotient
2. descending powers
3. zero

Objective A
1. $x+5$
3. $3x-4$
5. $2x^2+11x+15$
7. $x^2-5x+25$
9. $x^3-10x^2+100x-1000$
11. $x^3+4x^2-20x-48+\dfrac{5}{x-3}$
13. $12x^2+19x+5$
15. $4x^3+6x^2+8+\dfrac{6}{2x+3}$

Copyright © 2013 Pearson Education, Inc.

Answers to Worksheets for Classroom or Lab Practice

Chapter 7 ADDITION AND SUBTRACTION OF RATIONAL NUMBERS AND EXPRESSIONS

7.1 Addition and Subtraction of Rational Numbers and Expressions with Like Denominators

Getting Ready
1. 21
2. 18
3. 4
4. $\dfrac{38}{7}$
5. $2^2 \cdot 7$
6. $(x-4)(x-7)$
7. $x^2 + 9x - 7$
8. $x^2 - 2x - 21$
9. $\dfrac{2}{5}$
10. 3
11. $\dfrac{x+2}{x+3}$

Key Terms
1. common denominators
2. numerators
3. difference
4. improper fractions

Objective A
1. $\dfrac{4}{5}$
3. $\dfrac{1}{6}$
5. $\dfrac{5}{7}$
7. 2
9. $\dfrac{1}{15}$

Objective B
11. $8\dfrac{2}{3}$
13. $-\dfrac{1}{4}$

Objective C
15. $\dfrac{7}{x}$
17. $\dfrac{8y}{z^2}$
19. $\dfrac{7}{x+3}$
21. $-\dfrac{8n}{2-m}$
23. $\dfrac{8-v}{v+8}$
25. $\dfrac{x-3}{x+1}$
27. $\dfrac{a+7}{a-6}$
29. $\dfrac{3}{u-3}$

7.2 Least Common Multiple and Equivalent Rational Expressions

Getting Ready
1. $2^2 \cdot 3 \cdot 5$
2. $(3p+4)(3p-4)$
3. $(5x+1)(2x-3)$
4. 8
5. $5y^2$
6. $24x^4$
7. $12x^5 y^5$
8. $\dfrac{15}{35}$
9. $\dfrac{42m^3 n}{18m^6}$
10. $\dfrac{5y+10}{y^3 + y^2 - 2y}$

Key Terms
1. least common multiple
2. GCF
3. equivalent

Answers to Worksheets for Classroom or Lab Practice

4. Fundamental Property of Fractions

Objective A
1. 72
3. 54

Objective B
5. $x^6 y^3$
7. $8p^4 q^2$
9. $5(m+4)$
11. $(y+3)(y+10)(y-10)$

Objective C
13. 24
15. $8c$
17. $16xy^2$
19. $10m^3 p^2 q^3$

Objective D
21. $4y$
23. $7a-49$
25. $9y^2 - 27y$
27. $a^2 + a - 2$

7.3 The Least Common Denominator of Fractions and Rational Expressions

Getting Ready
1. 75
2. $24x^2 y^3$
3. $(x+1)(x-3)(x+4)$
4. $18x^3 y^5$
5. $(x-8)(x+4)$
6. $\dfrac{9}{24}$
7. $\dfrac{14x}{6x^4 y^4}$
8. $\dfrac{x^2 + 2x - 15}{x^3 + 2x^2 - 11x - 12}$

Key Terms
1. least common denominator
2. rational numbers

Objective A
1. $72, \dfrac{16}{72}, \dfrac{45}{72}$
3. $72, \dfrac{27}{72}, \dfrac{20}{72}$
5. $108, \dfrac{10}{108}, \dfrac{105}{108}$
7. $140, \dfrac{105}{140}, \dfrac{63}{140}, \dfrac{84}{140}$

Objective B
9. $ab, \dfrac{5b}{ab}, \dfrac{4a}{ab}$
11. $qv, \dfrac{pv}{qv}, \dfrac{uq}{qv}$
13. $xyz, \dfrac{9z}{xyz}, \dfrac{20x}{xyz}$
15. $x^2 y^4 z^3, \dfrac{8ay^3}{x^2 y^4 z^3}, \dfrac{7bxz}{x^2 y^4 z^3}$
17. $22b^4 c^3, \dfrac{8abc^2}{22b^4 c^3}, \dfrac{5t}{22b^4 c^3}$
19. $(c+1)(c+3), \dfrac{c+3}{(c+1)(c+3)}, \dfrac{2c+2}{(c+1)(c+3)}$
21. $28(m-3), \dfrac{7m}{28(m-3)}, \dfrac{8m}{28(m-3)}$
23. $(x+8)(x-8)(x-7), \dfrac{x^2 - 7x}{(x+8)(x-8)(x-7)}, \dfrac{4x^2 + 32x}{(x+8)(x-8)(x-7)}$

Copyright © 2013 Pearson Education, Inc.   481

Answers to Worksheets for Classroom or Lab Practice

25. $(w-3)(w+2)(w-4)$, $\dfrac{w^2+2w-24}{(w-3)(w+2)(w-4)}$, $\dfrac{w^2+w-2}{(w-3)(w+2)(w-4)}$

27. $(x-7)(x-2)(x+9)$, $\dfrac{x^2+10x+9}{(x-7)(x-2)(x+9)}$, $\dfrac{x^2+x-56}{(x-7)(x-2)(x+9)}$

29. $n^2(n+5)(n-4)$, $\dfrac{n^2+2n}{n^2(n+5)(n-4)}$, $\dfrac{n^2-10n+24}{n^2(n+5)(n-4)}$

7.4 Addition and Subtraction of Rational Numbers and Expressions with Unlike Denominators

Getting Ready
1. $\dfrac{29}{8}$
2. $\dfrac{19}{45}$
3. $\dfrac{6x+1}{6x}$
4. $\dfrac{2}{5}$
5. $(x-7)(2x+4)$
6. $(a-7)^2$
7. $\dfrac{21}{56}$ and $\dfrac{40}{56}$
8. $\dfrac{27}{45m}$ and $\dfrac{35}{45m}$
9. $\dfrac{2x+8}{6(x-3)}$ and $\dfrac{3x+15}{6(x-3)}$

Key Terms
1. least common denominator
2. trinomials

Objective A
1. $\dfrac{83}{72}$
3. $-\dfrac{64}{5}$
5. $\dfrac{3}{2}$
7. $\dfrac{57}{10}$
9. $\dfrac{220}{21}$ or $10\dfrac{10}{21}$
11. $\dfrac{67}{60}$
13. $\dfrac{691}{36}$ or $19\dfrac{7}{36}$
15. $\dfrac{23}{8}$ or $2\dfrac{7}{8}$

Objective B
17. $\dfrac{5y-x^2}{xy}$
19. $\dfrac{1}{18t}$
21. $\dfrac{3p+31}{20p}$
23. $\dfrac{-4r+50}{r(r-5)}$
25. $\dfrac{x+2}{(x-7)(x-6)}$
27. $\dfrac{28y^2+13y+2}{(3y-1)(2y+5)}$
29. $\dfrac{9t^2-13t}{(t+2)(t+3)(t-7)}$
31. $\dfrac{2y^2+18y+56}{(y+3)(y+8)(y+4)}$
33. $\dfrac{-19w-35}{(w+5)^2(w-1)}$

7.5 Complex Fractions

Getting Ready
1. $\dfrac{6}{7}$
2. $\dfrac{z^2y}{x^2}$
3. $\dfrac{4y}{x^2z^4}$
4. $\dfrac{y+4}{y}$
5. $(x+6)(x+9)$

Answers to Worksheets for Classroom or Lab Practice

6. a^2 7. $b^3 + 4b^2$ 8. $y^2 - 7y + 4$ 9. $\dfrac{39}{76}$ 10. $\dfrac{ac^3}{b}$

Key Terms
1. simplify   2. complex   3. one

Objective A
1. $\dfrac{2}{3}$ 3. $\dfrac{4}{3}$ 5. 8   7. $\dfrac{51}{2}$ 9. $\dfrac{54}{77}$ 11. $\dfrac{184}{411}$

Objective B
13. $\dfrac{3a}{2b}$ 15. $\dfrac{a}{bc}$ 17. $\dfrac{x}{y^2 z^2}$ 19. $\dfrac{3a}{24 + a}$
21. $\dfrac{p+3}{p-5}$ 23. $-\dfrac{5}{16}$ 25. $\dfrac{q+8}{q(q+9)}$

7.6 Solving Equations Containing Rational Numbers and Expressions

Getting Ready
1. 24   2. $12(x+1)$ 3. $(x+4)(x+5)(x-3)$ 4. $-19x - 42$
5. 5   6. $x = 3$ 7. $x = -3$ 8. $x = -5, 6$

Key Terms
1. multiplication property of equality   2. LCD   3. extraneous

Objective A
1. $\dfrac{11}{3}$ 3. $\dfrac{7}{6}$ 5. $\dfrac{20}{21}$ 7. $\dfrac{25}{11}$ 9. 3

Objective B
11. 10   13. $\dfrac{5}{12}$ 15. 4   17. 4, 1   19. -5
21. $-8, 1$ 23. $-\dfrac{19}{6}$ 25. $\varnothing$

7.7 Applications with Rational Expressions

Getting Ready
1. $x = 36$ 2. $x = -625$ 3. $x = -18$ 4. $x = -\dfrac{9}{11}$ or 6
5. $x, x+2$ 6. $x, x+2, x+4$

Copyright © 2013 Pearson Education, Inc.

Answers to Worksheets for Classroom or Lab Practice

Key Terms
1. $x+1$
2. $x+2$
3. $\dfrac{1}{x}$
4. x
5. rt

Objective A
1. 12, 18
3. 5, 7
5. $36,000
7. 5, 20

Objective B
9. $\dfrac{24}{7}$ hr = $3\dfrac{3}{7}$ hr ≈ 3 hr 26 min
11. $\dfrac{280}{3}$ min or $93\dfrac{1}{3}$ min
13. 60 hr

Objective C
15. 3 hr
17. Camela: 65 mph; Sarah: 60 mph
19. 403 mph

Chapter 8   RATIOS, PERCENTS, AND APPLICATIONS

8.1   Ratios and Rates

Getting Ready
1. $2^3 \cdot 3^2$
2. $\dfrac{1}{3}$
3. $\dfrac{8}{81}$
4. $\dfrac{1}{4}$

Key Terms
1. common factors
2. rates
3. ratio
4. unit price

Objective B
1. 1 to 4
3. 3:19
5. $\dfrac{40}{57}$
7. $\dfrac{9}{8}$
9. $\dfrac{1}{6}$

Objective C
11. $\dfrac{3}{4}$
13. $\dfrac{9}{8}$

Objective D
15. 55 mph
17. $31,250 per person

Objective E
19. $0.018
21. $2.75
23. $0.036

8.2   Proportions

Getting Ready
1. 378
2. $x = 18$
3. $x = 1$
4. $x = -5$ or 6
5. 18.3

Answers to Worksheets for Classroom or Lab Practice

6. 32   7. 36   8. $t = \dfrac{d}{r}$

Key Terms
1. cross products   2. proportion

Objective A
1. yes   3. no

Objective B
5. $x = 14$ 7. $a = 605$ 9. $x = 35$ 11. $m = 18$
13. $p = 4$ 15. $x = 10$ 17. $x = 2, 3$ 19. $x = 2$

Objective C
21. 540 miles   23. $36\dfrac{2}{3}$ ft   25. 3.75 bags

8.3 Percent

Getting Ready
1. .0014   2. $\dfrac{39}{11}$ 3. $\dfrac{19}{500}$ 4. .118   5. 73.33   6. 2.6
7. $11\dfrac{1}{2}$

Key Terms
1. percent   2. .01   3. $\dfrac{1}{100}$

Objective A
1. $\dfrac{9}{50}$ 3. $\dfrac{17}{10}$ 5. $\dfrac{13}{400}$

Objective B
7. .81   9. .007   11. 2.41   13. .0506   15. .104

Objective C
17. 47%   19. .07%   21. 498%   23. $44\dfrac{4}{9}\%$

Objective D
25. 20%   27. $37\dfrac{1}{2}\%$ 29. 114.3%

Answers to Worksheets for Classroom or Lab Practice

Objective E
31. 39%   33. 70.7%

8.4 Applications of Percent

Getting Ready
1. $\frac{7}{10}$ 2. .08   3. 4176   4. 4500   5. $x = 80$
6. $y = 65$ 7. $P = 108$ 8. $A = 180$

Key Terms
1. amount   2. equals   3. base   4. multiplication
5. number of percent

Objective A
1. What number is 14% of 38?   3. 22% of what number is 7?

Objective B
5. 48   7. 17.5   9. 400%   11. 600

Objective C
13. 13.5   15. 20%   17. 48   19. 28.2

Objective D
21. $33\frac{1}{3}\%$ 23. 10.24 oz

8.5 Further Applications of Percent

Getting Ready
1. $P = 200$ 2. .72   3. $\frac{6}{25}$ 4. 7.7   5. 450
6. $p = 700$ 7. 131.4   8. 60%

Key Terms
1. compound   2. sales tax   3. commission   4. purchase price
5. discount   6. total cost   7. principal   8. sales

Objective A
1. $2.70   3. $340

Objective B
5. $24   7. $1180; 16.9%

Answers to Worksheets for Classroom or Lab Practice

Objective C
9. 12%

Objective D
11. $96   13. 9 yr   15. $25.89   17. $109.38

Objective E
19. $1285.47

Objective F
21. 25%   23. 54.4°   25. 21.4%

Chapter 9 SYSTEMS OF LINEAR EQUATIONS

9.1 Defining Linear Systems and Solving by Graphing

Getting Ready
1. yes   2. no   3. $x = 29$ 4. $y = -1$ 5. no   6. yes

Key Terms
1. consistent   2. inconsistent   3. dependent   4. solution
5. system   6. coordinates

Objective A
1. solution   3. not a solution   5. solution

Objective B
7a. consistent b. 1 c. (3,1)   9a. inconsistent b. 0
11. (1,5)   13. (0,6)

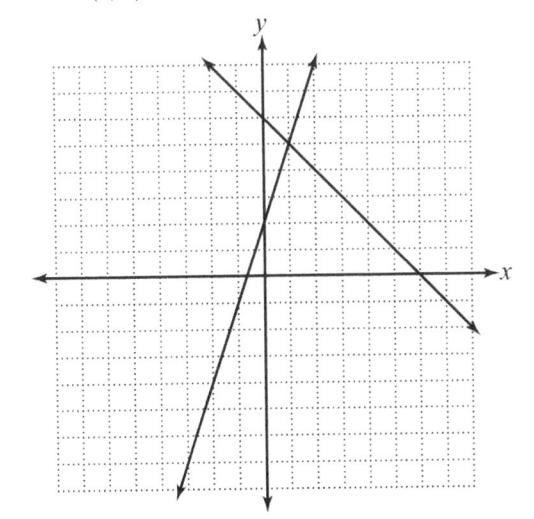

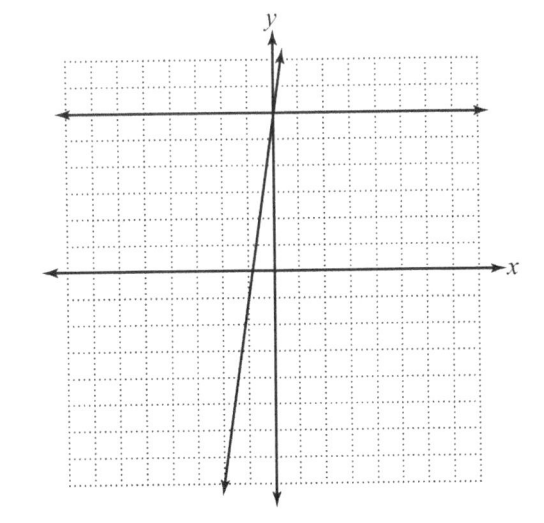

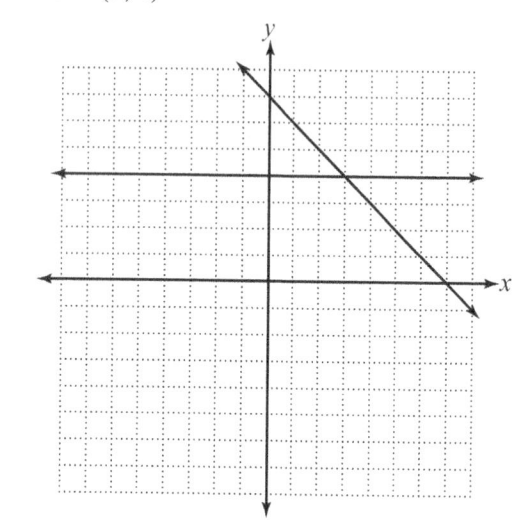

Answers to Worksheets for Classroom or Lab Practice

15. (3, 4)

17. (0, 0)

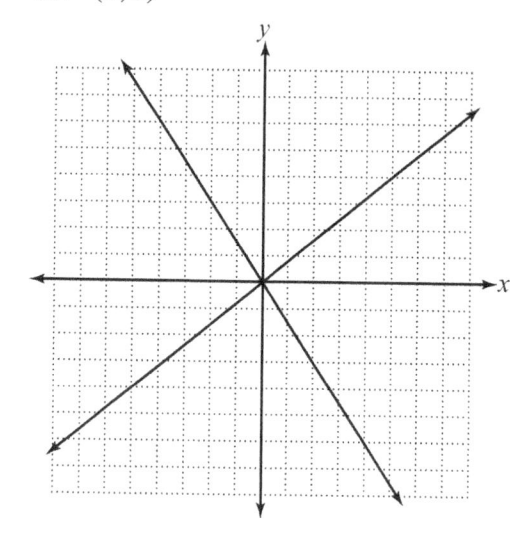

Objective C
19. dependent   21. inconsistent

9.2 Solving Systems of Linear Equations by Using Elimination by Addition

Getting Ready
1. 2
2. 16
3. −27
4. −3
5. $2x+7$
6. $-a+4$
7. yes
8. −4
9. $6x+y=4$
10. $x=-7$
11. $35x-7y$

Key Terms
1. Addition of Equals   2. additive inverses   3. LCD

Objective A
1. (5, 1)   3. (0, 3)   5. (7, −6)   7. (8, 0)   9. (−1, 12)
11. $\left(\dfrac{1}{2}, 6\right)$ 13. $\left(\dfrac{4}{5}, -\dfrac{1}{6}\right)$ 15. (0, 8)

Objective B
17. dependent   19. inconsistent   21. consistent; (0, 8)

9.3 Solving Systems of Linear Equations by Using Substitution

Getting Ready
1. $y=3x+8$
2. $y=22$
3. no
4. 3
5. 15
6. 12
7. −6
8. −12
9. dependent

Key Terms
1. inconsistent   2. dependent   3. ± 1 4. substitution

488   Copyright © 2013 Pearson Education, Inc.

Answers to Worksheets for Classroom or Lab Practice

Objective A
1. $(1, -2)$
3. $(-5, -6)$
5. $(0, 14)$
7. $(-8, -1)$
9. $(-2, 2)$
11. $(-7, -6)$
13. $\left(\dfrac{1}{2}, 2\right)$
15. dependent
17. inconsistent
19. dependent

9.4 Solving Application Problems by Using Systems of Equations

Getting Ready
1. $(5, -2)$
2. $(-1, 3)$
3. 8
4. 72
5. .05
6. .76
7. .18

Key Terms
1. variables
2. left
3. time

Objective A
1. German: 49 million; Russian: 3 million
3. 29, 83

Objective B
5. CD: $12; DVD: $16
7. shirt: $9; pants: $16

Objective C
9. plane: 400 mph; wind: 55 mph
11. Betsy: 50 mph; Aaron: 65 mph

Objective D
13. $26,000
15. amount at 4%: $80,000; amount at 8%: $40,000

Objective E
17. 610 tickets
19. 6% solution: 14 oz; 18% solution: 16 oz

Answers to Worksheets for Classroom or Lab Practice

9.5 Systems of Linear Inequalities

Getting Ready

1.

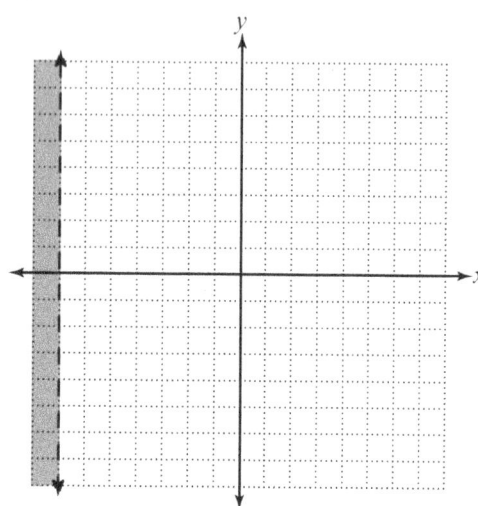

2.

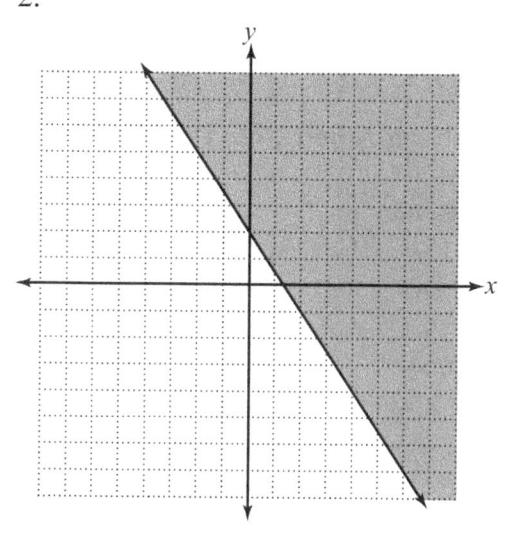

3.

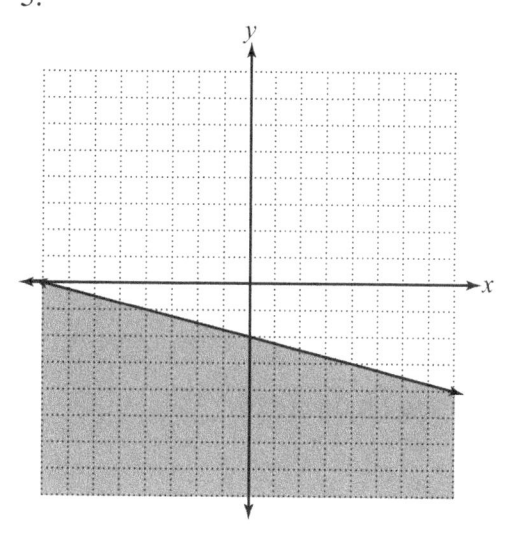

Key Terms
1. system
2. solid
3. dashed

Answers to Worksheets for Classroom or Lab Practice

Objective A
1. 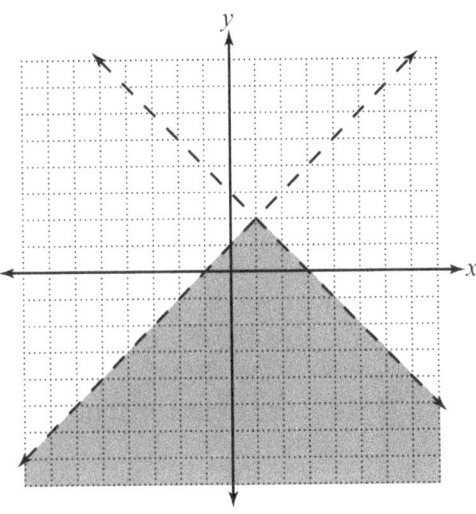
3. 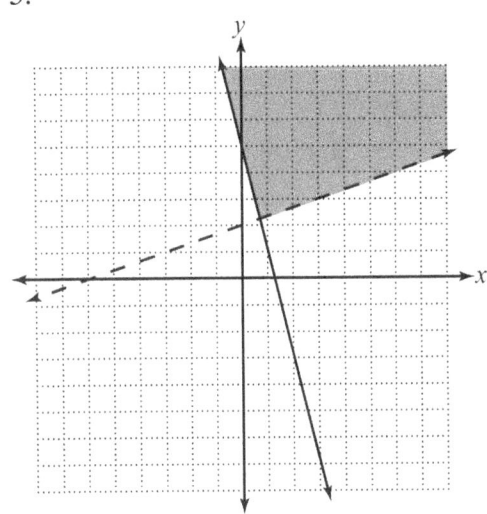
5. 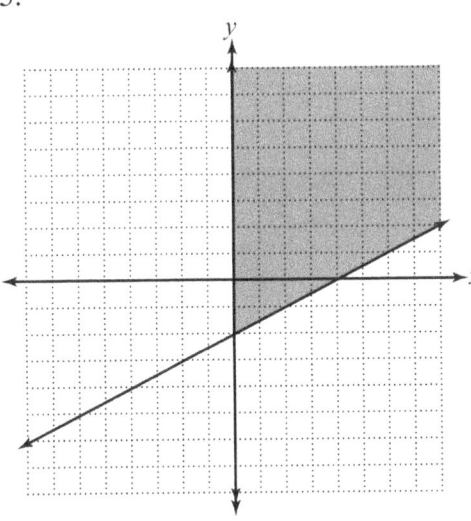
7. 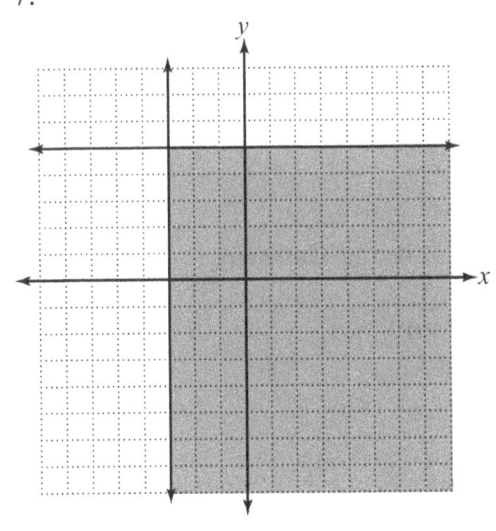
9.

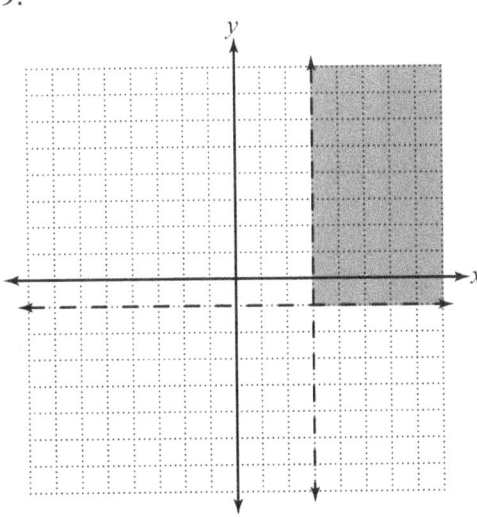

Copyright © 2013 Pearson Education, Inc.   491

Answers to Worksheets for Classroom or Lab Practice

Chapter 10   ROOTS AND RADICALS

10.1 Defining and Finding Roots

Getting Ready
1. 16   2. 243   3. −27   4. 216   5. x^{12} 6. a^8

Key Terms
1. irrational   2. square root   3. root index   4. radicand
5. radical   6. n^{th} root   7. perfect squares   8. $\sqrt{}$

Objective A
1. 6   3. ±11   5. −15   7. does not exist as a real number
9. 5   11. −4   13. −2   15. 1

Objectives B, C
17. irrational, 5.292   19. rational, .5   21. irrational, 3.302
23. rational, 28

Objective D
25. x^4 27. d^4 29. d 31. $x+7$ 33. $y+5$

10.2 Simplifying Radicals

Getting Ready
1. 96   2. 140   3. 4   4. 5   5. $2^2 \cdot 3 \cdot 7$ 6. 5
7. 7   8. 4   9. 6   10. x^{14} 11. $a^4 b^6$

Key Terms
1. $\sqrt[n]{a} \cdot \sqrt[n]{b} = \sqrt[n]{a \cdot b}$ 2. simplified form

Objectives A, B
1. $2\sqrt{6}$ 3. $7\sqrt{6}$ 5. $35\sqrt{7}$ 7. $y^9 \sqrt{y}$
9. $xy^3 z^5 \sqrt{yz}$ 11. $p^3 q^3 \sqrt{42pq}$ 13. $4t^5 w\sqrt{3tw}$ 15. $63 m^{11} n^{13} \sqrt{n}$

Objective C
17. $2\sqrt[3]{17}$ 19. x^3 21. $m^8 \sqrt[3]{m^2}$ 23. $4p^2 q^3 \sqrt[3]{pq}$
25. $4b^3 c\sqrt[3]{3c^2}$ 27. $p^9 q^{10} \sqrt[3]{q}$ 29. $50\sqrt[3]{4}$

Answers to Worksheets for Classroom or Lab Practice

10.3 Products and Quotients of Radicals

Getting Ready
1. 72
2. $21x^8$
3. 112
4. $16xy^4$
5. 9
6. $a^4 b$
7. $4x^2\sqrt{5x}$
8. $15ab^3\sqrt{2a}$

Key Terms
1. $\sqrt[n]{\dfrac{a}{b}}$
2. $\sqrt[n]{a \cdot b}$
3. a

Objective A
1. $\sqrt{55}$
3. $\sqrt{51}$
5. $2\sqrt{3a}$
7. x^3
9. z
11. $192\sqrt{3}$
13. $p^6\sqrt{p}$
15. $6x^9 y^7$
17. x^5
19. z
21. $6p^7$

Objective B
23. $\dfrac{5}{4}$
25. 8
27. 2
29. $ab^2\sqrt{a}$
31. $p^3\sqrt{3p}$

Objective C
33. $\dfrac{2\sqrt{7}}{11}$
35. $\dfrac{3\sqrt{14x}}{25}$

10.4 Addition, Subtraction, and Mixed Operations with Radicals

Getting Ready
1. x
2. $-4x^3 y^2$
3. $20\sqrt{3}$
4. $5\sqrt{6}$
5. $\sqrt{15}$
6. $24x + 42x^2$
7. $4x^2 - 20xy - 56y^2$
8. $9x^2 + 24xy + 16y^2$
9. $x^2 - 16y^2$
10. $1 + 2x$

Key Terms
1. conjugates
2. like radical expressions
3. coefficients

Objective A
1. $23\sqrt{5}$
3. $20\sqrt{17}$
5. $5a\sqrt{b}$

Objective B
7. $34\sqrt{6}$
9. $15\sqrt{7}$
11. $9\sqrt{6}$
13. $65x\sqrt{3x}$
15. $-11p^2\sqrt{5}$
17. $97\sqrt{5}$
18.

Copyright © 2013 Pearson Education, Inc.

Answers to Worksheets for Classroom or Lab Practice

Objective C
19. $-22\sqrt{2}$
21. $2\sqrt{5}$

Objective D
23. $6\sqrt{7}-7$
25. $48\sqrt{15}+160$
27. $72+17\sqrt{x}+x$
29. $a+3\sqrt{ab}-18b$
31. $24\sqrt{15}-3\sqrt{35}+48\sqrt{2}-2\sqrt{42}$
33. $31-10\sqrt{6}$
35. $3+\dfrac{\sqrt{6}}{2}$
37. 34
39. $49-x$

10.5 Rationalizing the Denominator

Getting Ready
1. $8x^3$
2. $2x$
3. $56-7\sqrt{3}$
4. 22
5. $\sqrt{a}-\sqrt{ab}+\sqrt{b}-b$
6. $6x^3$
7. $4x^3$
8. $2^4 \cdot b^4$
9. $\dfrac{4+2\sqrt{3}}{5}$

Key Terms
1. rationalizing the denominator
2. one
3. conjugates

Objective A
1. $\dfrac{9\sqrt{11}}{11}$
3. $-6\sqrt{3}$
5. $\dfrac{16\sqrt{15}}{15}$
7. $2\sqrt{w}$
9. $\dfrac{2\sqrt{p}}{p}$
11. $\dfrac{\sqrt{30}}{12}$

Objective B
13. $\dfrac{7\sqrt{2}}{2}$
15. $\dfrac{9\sqrt{5}}{5}$
17. $\dfrac{4w^3\sqrt{wt}}{t}$
19. $\dfrac{5a^2\sqrt{ab}}{b}$
21. $\dfrac{\sqrt[3]{175}}{5}$
23. $\dfrac{\sqrt[3]{63}}{3}$
25. $\dfrac{\sqrt[3]{20nm^2}}{5m}$
27. $\dfrac{\sqrt[3]{99ab^2}}{9b}$

Objective C
29. $\dfrac{7\sqrt{3}-\sqrt{15}}{44}$
31. $\dfrac{14+2\sqrt{2}}{47}$
33. $\dfrac{3\sqrt{2}-12\sqrt{3}-\sqrt{6}+12}{2}$
35. $\dfrac{m+2\sqrt{mn}+n}{m-n}$

10.6 Solving Equations with Radicals

Getting Ready
1. $4x+5$
2. $38-12\sqrt{x+2}+x$
3. $x=4$
4. $x=-5,\ -7$

Answers to Worksheets for Classroom or Lab Practice

5. $\dfrac{4}{3}$

Key Terms
1. extraneous   2. square   3. isolate

Objective A
1. 4   3. ∅   5. ∅   7. 8   9. 4   11. 6
13. 0, 7   15. 2

Objective B
17. 17   19. 3, −1   21. ∅   23. 1

10.7 Pythagorean Theorem

Getting Ready
1. 150   2. 18   3. $a^2 = 45$ 4. $2\sqrt{19}$

Key Terms
1. right   2. Pythagorean Theorem   3. legs
4. hypotenuse   5. diagonal

Objective A
1. 1. $\sqrt{58}$ in.   3. $2\sqrt{21}$ cm   5. $4\sqrt{13}$ ft   7. $2\sqrt{13}$ in.
9. 28 ft

Objective B
11. 70 cm   13. $2\sqrt{14}$ in.

Objective C
15. $6\sqrt{5}$ cm   17. $2\sqrt{38}$ ft

Objective D
19. 67.7 yd   21. 26.9 mi

Chapter 11   SOLVING QUADRATIC EQUATIONS

11.1 Solving Incomplete Quadratic Equations

Getting Ready
1. $2x(8x-1)$ 2. $x = \dfrac{8}{3}$ 3. 5   4. $4\sqrt{3}$ 5. $\dfrac{\sqrt{10}}{5}$
6. 125

Answers to Worksheets for Classroom or Lab Practice

Key Terms
1. positive
2. incomplete quadratic equation
3. factorable

Objective A
1. $0, -6$
3. $0, 4$
5. $0, \dfrac{25}{4}$
7. $0, 12$

Objective B
9. ± 4
11. no real-number solutions
13. ± 14
15. $\pm\sqrt{30}$
17. ± 8
19. $\pm\sqrt{3}$

Objective C
21. $0, 10$
23. $6, -3$
25. $-7 \pm \sqrt{15}$
27. $-9 \pm \sqrt{3}$
29. $\dfrac{1 \pm \sqrt{42}}{2}$

Objective D
31. $2, -12$
33. 15 ft

11.2 Solving Quadratic Equations by Completing the Square

Getting Ready
1. $12 + 6\sqrt{3}$
2. $x^2 + 5x + \dfrac{25}{4}$
3. $\dfrac{64}{49}$
4. $\left(a + \dfrac{9}{2}\right)^2$
5. $\dfrac{21}{44}$
6. $x = -6, -1$
7. $x = 3 - \sqrt{3}, 3 + \sqrt{3}$
8. $\dfrac{5}{4}$
9. $4\sqrt{5}$
10. $\dfrac{4\sqrt{3} - 5}{3}$
11. $\dfrac{-\sqrt{5} - 6}{2}$
12. $\dfrac{\sqrt{15}}{5}$

Key Terms
1. perfect square trinomial
2. $\dfrac{1}{2}b$
3. completing the square
4. 1

Objective A
1. $9; (x-3)^2$
3. $36; (x-6)^2$
5. $\dfrac{121}{4}; \left(q + \dfrac{11}{2}\right)^2$
7. $\dfrac{1}{121}; \left(p - \dfrac{1}{11}\right)^2$

Answers to Worksheets for Classroom or Lab Practice

Objective B
9. $-2, -10$
11. $\dfrac{-5 \pm 3\sqrt{5}}{2}$
13. $-4 \pm \sqrt{15}$
15. $\dfrac{1}{2}, -1$
17. no real-number solutions
19. $\dfrac{-1 \pm \sqrt{31}}{5}$

11.3 Solving Quadratic Equations by the Quadratic Formula

Getting Ready
1. $x = -4 + 3\sqrt{2}, -4 - 3\sqrt{2}$
2. $\dfrac{2+\sqrt{2}}{2}, \dfrac{2-\sqrt{2}}{2}$
3. $\dfrac{4}{3}, 1$
4. $5\sqrt{14}$
5. $-2 + \sqrt{17}$

Key Terms
1. coefficients
2. quadratic equation
3. Quadratic Formula
4. $ax^2 + bx + c = 0$

Objective A
1. no real-number solutions
3. $2 \pm \sqrt{6}$
5. $\dfrac{-4 \pm \sqrt{26}}{2}$
7. $1, \dfrac{1}{5}$
9. $-\dfrac{3}{2}$
11. $\dfrac{9}{2}, -1$
13. $\dfrac{-5 \pm \sqrt{15}}{2}$
15. $\dfrac{-3 \pm \sqrt{3}}{3}$
17. $\dfrac{-5 \pm \sqrt{13}}{3}$
19. $1, -\dfrac{5}{3}$

11.4 Quadratic Equations with Complex Solutions

Getting Ready
1. $4\sqrt{3}$
2. $2 + 3i$
3. $24 + 18i$
4. $23 + 22i$
5. 65
6. $\dfrac{7}{5} - \dfrac{4i}{5}$
7. $x = 3, 13$
8. $x = \dfrac{3+\sqrt{41}}{2}, \dfrac{3-\sqrt{41}}{2}$

Key Terms
1. complex conjugates
2. imaginary unit
3. real number
4. standard form
5. complex number

Objective A
1. $11i$
3. $2i\sqrt{2}$
5. $i\sqrt{17}$
7. $18 + 0i$
9. $0 + 12i$
11. $4\sqrt{2} + 0i$
13. $5 + 5i$

Objective B
15. $24 - 3i$
17. $4 - 6i$
19. $-24 + 5i$
21. $0 + 26i$

Answers to Worksheets for Classroom or Lab Practice

Objective C

23. $-18+14i$ 25. $6+4i$ 27. $65+0i$ 29. $\dfrac{1}{2}+\dfrac{3}{2}i$

31. $\dfrac{7}{5}-\dfrac{16}{5}i$

Objective D

33. $3\pm 7i$ 35. $3\pm 2i$ 37. $\dfrac{1\pm i\sqrt{3}}{2}$

11.5 Applications Involving Quadratic Equations

Getting Ready

1. $x=\dfrac{7}{3}$ 2. $x=-7, 6$ 3. $x=-5, 4$ 4. 3m, 4m
5. $x, x+2, x+4$

Key Terms

1. physical conditions 2. original problem

Objectives A–F

1. $-10, -9$ or $9, 10$ 3. 5 ft 5. 5 mph 7. 12 min, 8 min
9. 106 ft 11. Kyle: 9 days; Amanda: 18 days 13. 8, 10 or $-10, -8$
15. 6 in. by 13 in.